Novel Group Theoretical Methods for Electron Structure Theory

Victor G. Yarzhemsky

Novel Group Theoretical Methods for Electron Structure Theory

Springer

Victor G. Yarzhemsky
NS Kurnakov Institute of General
and Inorganic Chemistry of RAS
Moscow, Russia

ISBN 978-3-031-72801-3 ISBN 978-3-031-72802-0 (eBook)
https://doi.org/10.1007/978-3-031-72802-0

This Springer imprint is published by the registered company Springer Nature Switzerland AG
The registered company address is: Gewerbestrasse 11, 6330 Cham, Switzerland

Preface

Group theory makes it possible to label the states in atoms, molecules and solids by quantum numbers to construct invariant basis sets, and to obtain selection rules for transitions and interactions. In the present book, the main problems of group theory in the theory of electron structure, namely construction of invariant basis sets, finding normal vibrations in molecules and calculations of selection rules in solids, are solved by unified way making use of induced representation method. As a result, standard techniques for molecular orbital construction and estimation of normal vibration in molecules are significantly simplified. Recent progress in topological superconducting materials revealed a complicated structure of their order parameter, which is also may be described group theoretically. In this book, a new group-theoretical approach based on the theory of induced representations for two-electron states is presented. The main advantage of this approach is the ability to determine the dependence of electron pair symmetry on the symmetry of one-electron state and on non-symmorphic crystal structure. Two-electron states with zero total momentum are connected with Cooper pairs in topological superconductors. Two main problems are considered. Group-theoretical analysis of nodal structure of two-electron states in a space-group symmetry is developed. Also topological phase of the Cooper pair is connected with irreducible representation and the dependence of topological nodal structure on additional quantum numbers and time-reversal symmetry is considered. This part of the theory was developed by the author and appeared so far in journals only.

Different chapters of the book have a different focus on theoretical approaches and may be of interest to different groups of readers.

Chapter 1 gives a brief presentation of the foundations of group theory using examples of symmetric molecules and their wavefunctions. To describe the spin of an electron in a crystal field, instead of the standard approach of double groups, the method of projective representations is used, which allows one to consider the spatial and spin parts of the wavefunction in a unified way. The construction of two-electron wavefunctions in poit-group symmetry is considered, which lays the foundation for studying the many-electron theory.

Chapter 2 gives a brief presentation of the theory of corepresentations of magnetic groups and contains the examples of the application of the theory to the study of the symmetry of wavefunctions. Thus, Chaps. 1 and 2 contain the foundations of the theory of finite groups and its applications in the theory of electronic structure. Chapters 1 and 2 are introductory for the two last chapters and are mainly aimed at students studying the group-theoretical methods in qantum mechanics and in the theory of electronic structure.

In Chap. 3, on the basis of the theory of induced representations, three main problems of the theory of electronic structure are considered: the construction of multicenter wavefunctions of clusters, the determination of the symmetry of normal vibrations of molecules and the construction of irreducible representations of space groups. The third chapter may be of interest to specialists in the field of electronic structure.

In Chap. 4, the space-group approach to the wavefunction of a Cooper pair, proposed for the first time by author, is presented. In connection with the topicality of the problem of searching for new topological superconductors, this approach was also developed by other authors, the results of which are also reflected in the Chap. 4. This chapter also presents the author's new approach, which makes it possible to combine group-theoretical and topological approaches. Chapter 4 is of interest to specialists in the field of topological materials.

Moscow, Russia Victor G. Yarzhemsky

Introduction

Every process in physics is governed by selection rules that are the consequence of symmetry requirements. The beauty and strength of group theory resides in the transformation of many complex symmetry operations into a very simple linear algebra M. Dresselhouse et al. [1].

The above words of Mildred Dresselhouse, the author of the modern "unique textbook" on the application of group theory in condensed matter physics [1], perfectly reflect the role of group theory in physics and the theory of electron structure of condensed matter. In addition, group theory makes it possible to construct wavefunction basis sets and to assign physical quantum numbers to these sets.

The group-theoretical methods in physics of molecules and solids set forth in the collection of books that have already become classical [3–14], make it possible to find the wavefunctions of atoms in a crystal field, symmetrized linear combinations of orbitals in molecules (molecular orbitals or MO) and normal vibrations. Some of these topics are included in textbooks on theoretical physics [15]. Induced representation method has been both useful and powerful. It has been recognized [16–18] that the application of the method of induced representations, described in the books of Bradley and Cracknell [19] and Altmann [20], namely a Frobenius reciprocity theorem [22, 23], allows one to quickly find normal vibrations and MO without consideration of the whole basis set as in standard approaches [9, 11–13, 15, 21]. General theory of reduction factors for induced representations was developed by Haase, Buttler and Dirl [24, 25]. When constructing MOs or normal vibration of large symmetric clusters, the problem of physical labeling of repeating IRs (irreducible representations) arises. The transitivity of induction theorem [19, 20] makes it possible to label repeating IRs by IRs of a physically distinguished intermediate subgroup and to construct corresponding basis sets [26–30]. The application of the methods of induced representations is especially required for MO and normal vibration analysis in large symmetrical clusters consisting of several orbits of atoms. Similar approaches, i.e. construction of the wavefunctions on the basis of atomic orbitals were applied to solids [17, 31–34] and are called band representations.

Mackey theorems [35, 36] and their new forms for the space groups, obtained by Bradley and Davies [37, 38] have proven to be very effective for construction of singlet and tripled wavefunctions of electron pairs in symmetrical clusters and crystal solids [39–41]. Currently, this approach is recognized as the most general group-theoretical approach for gap classification in non-symmorphic superconductors [42–47]. Mackey theorem was used for unique labeling of many-electron states in multicentre systems [48]. Recently, the space-group approach to the wavefunction of a Cooper was further developed by making use of magnetic-group symmetry, transitivity of induction theorem and was connected with phase winding of a pair [49–51] and to construct four-electron states in solids [52, 53]. These new group-theoretical results are outlined in Chaps. 3 and 4 on the basis of standard group-theoretical approaches, which are considered in Chaps. 1 and 2.

Contents

Chapter 1
Point Groups and Wavefunctions in Molecules

Weyl, Wigner and van der Wärden developed the theory in a style of great generality and mathematical polish, which delayed for about 25 years its acceptance as a working tool by the majority of physicists, although the chemists during the same period found no great difficulty in adapting it to their needs.

F.J. Dyson [54]

Abstract The basic definitions and theorems of group theory, which are used in the standard theory of the electronic structure of atoms in a crystal field and molecules, are briefly considered in this chapter. Particular attention is paid to the methods for constructing two-electron states and wavefunctions in multicenter systems. To consider the spin parts of the wavefunctions, instead of the standard approach of double groups, the method of projective representations is used. The application of the approaches is illustrated by many examples of the construction of electron states in point group symmetry.

1.1 Groups, Subgroups, Left Cosets and Orbits

The symmetry properties of quantum systems are best described using the mathematical apparatus of group theory. According to definition, group G is a set of elements p, q, r, s of an arbitrary nature, for which an operation called multiplication is defined, i.e. each two elements r and s are associated with an element p, called their product:

$$p = sr \tag{1.1.1}$$

which is also an element of group G. When writing the multiplication of group elements, as well as when writing the usual multiplication operation, the multiplication sign is usually omitted. By definition a group must contain an identity E, which has the property:

$$Ep = pE = p \tag{1.1.2}$$

V. G. Yarzhemsky, *Novel Group Theoretical Methods for Electron Structure Theory*,
https://doi.org/10.1007/978-3-031-72802-0_1

By definition for each element p, the group must also contain the inverse element p^{-1}, which has the property:

$$p^{-1}p = pp^{-1} = E \tag{1.1.3}$$

The multiplication operation must satisfy the requirement of associativity:

$$p(qr) = (pq)r \tag{1.1.4}$$

In general, multiplication is not commutative:

$$pq \neq qp \tag{1.1.5}$$

The commutative group, i.e. the group in which the product of two elements does not depend on their order is called commutative or Abelian group. The number of the elements of a finite group G is called the order of the group and is denoted by $|G|$ throughout.

Quantum systems have different types of symmetry, which can be divided into two types—spatial symmetry and internal symmetry. The symmetry operation is understood as an operation that does not change the Hamiltonian of the system. The symmetry group of an atom is a group $\mathsf{O}(3)$, i.e. a group of pure rotations of the three-dimensional space $\mathsf{O}^{+}(3)$ supplemented by the inversion I. The spatial symmetry of a molecule consists of rotations through finite angles, inversion and mirror reflections. Such symmetry groups are called point groups. The crystal symmetry consists of operations of a point group α, of the translation subgroup T and of some point operations β associated with translations τ_β, that are not included in the translation subgroup T. The correct wavefunctions for the initial approximation should be constructed taking this geometric symmetry into account. In addition, electron wavefunctions have internal symmetry, in particular, symmetry with respect to permutations of coordinates in a many-electron system and time-reversal symmetry. All these operations satisfy the definitions of group theory and leave the Hamiltonian unchanged.

The example of commutative group is a group C_4– the group of four rotations about any axis. The four powers of the rotation element C_4, denoted as C_4, C_4^2, C_4^3 and $C_4^4 = E$ form an Abelian group C_4. If the axis is fixed, say $z-$axis, the axis symbol is added to the subscript of the element as C_{4z}. In order not to confuse the designations of the point group of rotations and its elements, we will write the first one in a different font. Also different fonts are used for group $\mathsf{O}(3)$ (group of rotations) and O_h (cubic group).

Let H be a subset of elements of a group G, which itself forms a group with respect to the multiplication law of G. This means that all possible products $h_i h_j$ of all pairs of elements from H and all inverse elements h_i^{-1} also belong to H. Also the identity element E of H is the same as identity element of G. Then H is called a subgroup of the group G. Note that the elements of the group G that do not belong to the subgroup H do not form a group, since they do not contain an identity element.

The symmetry group of a square, denoted as C_{4v}, contains eight elements—identity element E; elements C_{4z}, C_{2z} and C_{4z}^3—counterclockwise rotations about z-axis by angles $\pi/4$, $\pi/2$ and $3\pi/4$, respectively; and four reflections $\sigma_x, \sigma_y, \sigma_a, \sigma_b$ in planes perpendicular to the plane of the square (see Fig. 1.1).

The group structure can be defined as a multiplication table, where the element of the table is equal to the product of the element at the beginning of the row by the element at the top of the column. The structure of the C_{4v} group is presented in Table 1.1. The analysis of Table 1.1 shows that the elements of this group do not commute, for example, $C_{4z}\sigma_x = \sigma_a$, but $\sigma_x C_{4z} = \sigma_b$. It is also seen in Table 1.1 that in all rows and columns of the multiplication table of the group each element of the group occurs only once.

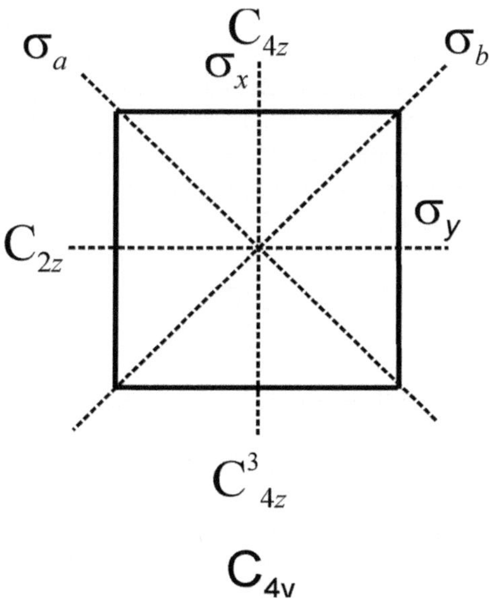

Fig. 1.1 Elements of C_{4v} group

Table 1.1 Multiplication table of the C_{4v} group (see Fig. 1.1 for notations). The product $g_i g_j$ is in the intersection of $i-$ row and $j-$ column

E	C_{4z}	C_{2z}	C_{4z}^3	σ_x	σ_y	σ_a	σ_b
C_{4z}	C_{2z}	C_{4z}^3	E	σ_a	σ_b	σ_y	σ_x
C_{2z}	C_{4z}^3	E	C_{4z}	σ_y	σ_x	σ_b	σ_a
C_{4z}^3	E	C_{4z}	C_{2z}	σ_b	σ_a	σ_x	σ_y
σ_x	σ_b	σ_y	σ_a	E	C_{2z}	C_{4z}^3	C_{4z}
σ_y	σ_a	σ_x	σ_b	C_{2z}	E	C_{4z}	C_{4z}^3
σ_a	σ_x	σ_b	σ_y	C_{4z}	C_{4z}^3	E	C_{2z}
σ_b	σ_y	σ_a	σ_x	C_{4z}^3	C_{4z}	C_{2z}	E

It can be obtained from Table 1.1 that group has three subgroups $C_4 = \{E, C_{4z}, C_{2z}, C_{4z}^3\}$, $C_{2v} = \{E, C_{2z}, \sigma_x, \sigma_y\}$ and $C'_{2v} = \{E, C_{2z}, \sigma_a, \sigma_b\}$.

A group can also be defined by its own generators. In the case of the group C_{4v}, it suffices to take the elements $g_1 = C_{4z}$ and $g_2 = \sigma_x$. An example of such a group representation is shown in Table 1.2.

Let H be a subgroup of G of order $|H|$ with elements $h_1, h_2, \ldots, h_m$. We choose some element s_2 not belonging to H and consider the collection of elements s_2h_1, $s_2h_2, \ldots, s_2h_m$, which is called the left coset and is denoted by s_2H. The element s_2 is called the left coset representative. Let us show that all elements of the left coset are distinct. In this case, it contains exactly $|H|$ elements. Assume the opposite, i.e. that there are two distinct elements $h_1 \neq h_2$ for which:

$$s_2h_1 = s_2h_2 \tag{1.1.6}$$

Multiplying from the left by s_2^{-1}, we obtain a contradiction with the original assumption, i.e. that $h_1 = h_2$. Since all elements of the left coset are distinct, the total number of elements in each left coset equals to $|H|$. Next, we choose an element s_3 that is not contained in either H or s_2H and compose another collection s_3H. Continuing the process, we obtain a decomposition of the group G into left cosets with respect to its subgroup H:

$$G = \sum_{i=1}^{p} s_i H \tag{1.1.7}$$

In this formula $s_1 = E$. Let us prove that the number of the left cosets is given by the relation:

$$p = |G|/|H| \tag{1.1.8}$$

Let us now prove that the left cosets either coincide or have no common elements. Assume the contrary, i.e. that $s_ih_1 = s_jh_2$. Multiplying on the right by h_1^{-1} we obtain that $s_i = s_jh_2h_1^{-1}$. This means that s_i belongs to the left coset s_jH and the left cosets formed by the elements s_i and s_j coincide, i.e. in fact, there is one left coset. Thus, since the number of elements in each of the left cosets is $|H|$ and the fact that the left cosets have no common elements, the relation (1.1.8) follows. It can be proven that any of the elements of the left coset can be chosen as a representative of this left coset.

Table 1.2 Elements of group C_{4v} represented as products of powers of its generators $g_1 = C_{4z}$ and $g_2 = \sigma_x$

g	$(g_1)^n\,(g_2)^k$	g	$(g_1)^n\,(g_2)^k$
E	E	σ_x	σ_x
C_{4z}	C_{4z}	σ_y	$(C_{4z})^2\,\sigma_x$
C_{2z}	$(C_{4z})^2$	σ_a	$(C_{4z})^3\,\sigma_x$
C_{4z}^3	$(C_{4z})^3$	σ_b	$C_{4z}\sigma_x$

For the subgroup H of G and any element s of group G such that $s \notin H$ consider the set of elements $H^s = sHs^{-1}$. A mapping from one group to another $H' = \varphi(H)$ is called an isomorphism if the laws of multiplication for the image and preimage are the same: $\varphi(h_i h_j) = \varphi(h_i)\varphi(h_j)$. It is easy to check that the collection of elements sHs^{-1} forms a group H^s isomorphic to H:

$$sh_i s^{-1} sh_j s^{-1} = sh_i h_j s^{-1} \tag{1.1.9}$$

If the transformed group sHs^{-1} coincides with the original group for any $s \in G$, then H is called an invariant subgroup.

The order (or index) of a subgroup H of group G is the ratio of the number of group elements to the number of subgroup elements $|G| / |H|$. Consider a subgroup H of group G of the order 2, i.e. the number of elements of the subgroup $|H|$ is half the number of elements of the group $|G|$. Let us choose an element s of the group G that does not belong to the subgroup H. We are going to prove that each subgroup of index 2 is an invariant subgroup, i.e.:

$$sh_i s^{-1} = h_j, \ h_i, \ h_j \in H \tag{1.1.10}$$

For any $s \notin H$ consider the set of elements sH, i.e. the left coset with respect to the subgroup H. Since H is a subgroup of order 2, $|H|$ elements of the group H and the same number of elements of the left coset sH exhaust all elements of the whole group G:

$$G = sH + H \tag{1.1.11}$$

Since the elements of the right coset together with the group H also exhaust the entire group, the set of elements of the right coset Hs coincides with the set of elements of the left coset sH. Then, expressing the common element $g \notin H$ as $sh_i = h_j s$, and multiplying by s^{-1} to the right we obtain the subgroup invariance condition:

$$sh_i s^{-1} = h_j \tag{1.1.12}$$

It also follows from the obtained properties of decomposition into left cosets that the order of a subgroup is a divisor of the order of the group. Subgroups of the order 2 are important in the theory of magnetic groups.

In addition to the properties of the elements of a group with respect to the operation of multiplication, the properties of the sets on which the elements of the group act are also essential. Consider a group G acting on a set X of elements $\{x, \ y...\}$. Two elements of this set x and y are called G—equivalent if there is such an element of the group s_i that:

$$s_i x = y \tag{1.1.13}$$

The classes of $G-$ equivalent elements of the set X are called the orbits of X with respect to the group G. The orbits are pairwise disjoint and their union coincides with the set X.

Let G be a point group and X be a set of points in a space lying on a sphere centered at the origin. For cubic point groups, it is convenient to take the surface of a cube instead of a sphere, and for hexagonal point groups, the surface of a regular hexagonal prism centered at the origin. The type of the orbit is determined by the local symmetry group H, i.e. subgroup that leaves the given point x in place. The group H is called the centralizer of the point x. Let's construct the orbit of the point x. We choose an element s_i of the group G that does not belong to H. This element maps the point x to the point x_i. It is obvious that all elements of the group represented as $s_i H$ also map the point x to the point x_i. The element s_i is a representative of the left coset. As a representative of the left coset, one can choose any of its elements. Continuing the expansion into the left cosets, we see that each left coset s_j corresponds to an orbit point $x_j = s_j x$ and total number of the elements in orbit of point x with centralizer H equals to the number of elements in left coset decomposition of G with respect to H, i.e. equals to $|G| / |H|$.

Atoms in symmetrical cluster of symmetry group G are distributed among several orbits. Consider as a example SF_6 molecule shown in Fig. 1.2. This molecule has octahedral symmetry O_h, which consists of 48 elements, including rotations, inversion and mirror reflections. Let us denote A_1 the ligand located on the positive direction of the $z-$axis. The centralizer of A_1 (local symmetry group) is a group C_{4v}, consisting of eight elements. Counterclockwise rotations by $2\pi/3$ and $4\pi/3$ about the axis (111) move atom A_1 into positions A_2 and A_3, respectively. The elements I, IC_3 and IC_3^2 move A_1 into positions A_4, A_5 and A_6, respectively. It can be directly verified making use of multiplication tables [60], that the above considered elements transforming A_1 into all other positions are representatives in the decomposition of O_h into left cosets with respect to C_{4v}. Symmetry groups of atoms in the orbit

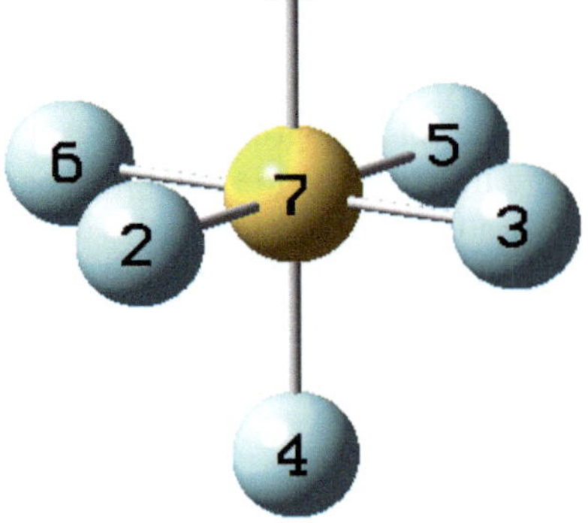

Fig. 1.2 SF_6 molecule of O_h symmetry

are isomorphic and may be obtained from symmetry group of the initial one by the transformation:

$$H^i = s_i H s_i^{-1} \tag{1.1.14}$$

In O_h symmetry atoms may be placed also on the threefold and twofold symmetry axis. In these cases, local groups are C_{3v} and C_{2v} of the orders 6 and 4 and orbits consist of 8 and 12 atoms, respectively. Eight atoms of the first orbit form a cube and 12 atoms of the second orbit form a cuboctahedron. Cluster Mo_8O_{12} shown in Fig. 1.3 consists of two orbits: an inner cubic orbit of 8 molybdenum atoms and outer cuboctahedron orbit of 12 oxygen atoms.

The icosahedral symmetry group I_h has fivefold and threefold symmetry axis, whose symmetry groups C_{5v} and C_{3v} are of the orders 10 and 6, respectively, and there are two corresponding orbits, namely, icosahedral orbit consisting of 12 atoms and dodecahedral orbit consisting of 20 atoms. Also the symmetry group in the center of icosahedron and dodecahedron in I_h group and corresponding orbit consists of one atom.

Figure 1.4 shows icosahedral cluster $Au_{12}W$ in which W atom is in the center and 12 Au atoms form an icosahedral orbit.

Figure 1.5 shows Au_{32} cluster in which the distances between the cluster center and atoms of both orbits Au_{12} and Au_{20} are approximately the same for two orbits.

Another possibility is shown in Fig. 1.6, where in a cluster $As@Ni_{12}@As_{20}$ of I_h symmetry inner orbit consists of 12 Ni atoms and outer orbit consists of 20 As atoms. It should be noted that in all cases the relation between the number of atoms in orbit and the number of elements in local symmetry group is given by formula (1.1.8).

In the case of symmetrical clusters, the main task of the group theory is to find symmetrized linear combinations of atomic orbital of all atoms of an orbit that are transformed by orthogonal matrices under the action of operations of symmetry group. Such combinations are found separately for each orbit. In the case

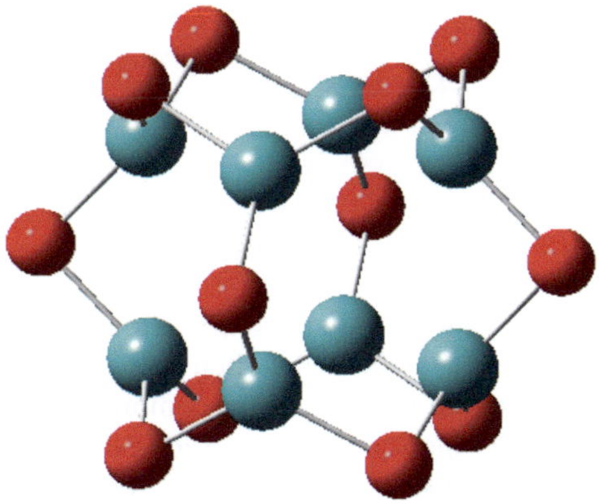

Fig. 1.3 Cluster Mo_8O_{12} of O_h symmetry

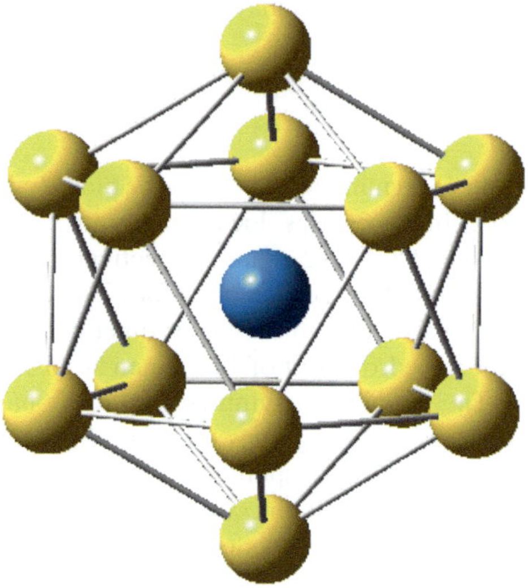

Fig. 1.4 Cluster $Au_{12}W$ of I_h symmetry

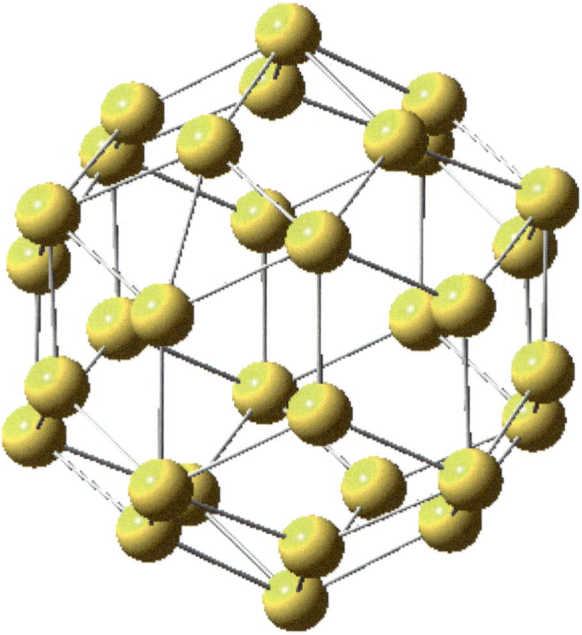

Fig. 1.5 Cluster Au_{32} of I_h symmetry consisting of 12 atomic and 20 atomic orbits of Au atoms

of an octahedron, such combinations are usually found by the method of projection operators [9]. However when going to I_h symmetry or space-group symmetry the construction of such linear combinations becomes very complex. The method of induced representations, considered in Chap. 3, allows one to analyze such orbitals without constructing them, using only existing tables of point group characters.

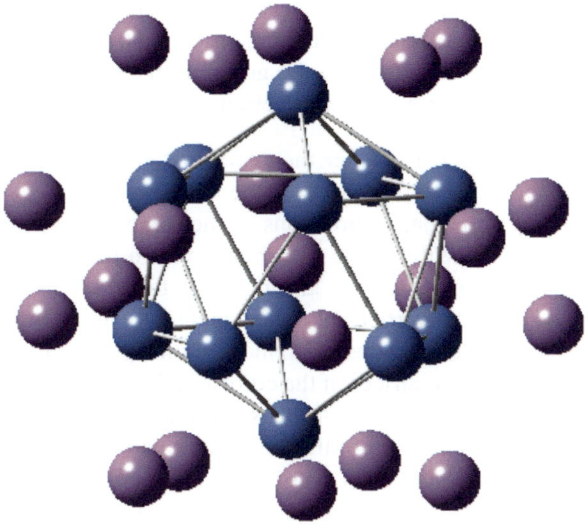

Fig. 1.6 Cluster As@$Ni_{12}As_{20}$ of I_h symmetry

1.2 Irreducible Representations and Characters

Let the elements r, s and t of the group G, connected by multiplication relation:

$$rs = t \tag{1.2.1}$$

be associated with n-dimensional matrices $D(r)$, $D(s)$ and $D(t)$, having the same multiplication relation:

$$D(r)D(s) = D(t), \ rs = t \tag{1.2.2}$$

Thus, the group G is mapped to the group of matrices with the same group law of multiplication. Then, the set of matrices $D(g)$ is called a representation of the group G. The identity element is associated with the identity $n-$dimensional matrix. The inverse element is associated with the inverse $n-$dimensional matrix $D(g^{-1}) = D^{-1}(g)$, so the determinants of the matrix $D(g)$ must be nonzero. It is important to note that the mapping $G \Longrightarrow D(g)$ is a homomorphism, i.e. it does not necessarily follow from $r \neq s$ that $D(r) \neq D(s)$. It is only required that each element g of G be uniquely associated with the matrix $D(g)$. At the same time, one matrix can be associated with several elements of the group. For example, each element of g can be associated with the same number— one. Such a representation is called an identity representation. In group theory, sums of diagonal elements of matrices of representations which are called the characters χ are very essential:

$$\chi\left[D(r)\right] = \sum_i D_{ii}(r) \tag{1.2.3}$$

Orbits of groups are spaces of representations. Under the action of the operations of the group, the elements of the orbit are transformed through each other according to the representations of the group. In the case of point groups, the elements of the orbit can be atomic wavefunctions or simply points of space.

Example 1.1 Consider the symmetry group O_h and place the points s_1, s_2 and s_3 on the positive directions of the coordinate axes z, x and y, and the points s_4, s_5 and s_6 on negative directions of the same axes (see Fig. 1.2) These points may be associated with atomic $s-$orbitals, which form an invariant space under the action of the elements of O_h group. Thus, these six $s-$orbitals s_1, s_2, s_3, s_4, s_5 and s_6 located at the vertices of the octahedron form an orbit of the group O_h. The actions of the elements of the group induce some permutation of these elements. It is easy to write out the matrices of the corresponding representation: these are six-dimensional permutation matrices of basis vectors, each row (column) of which contains one element. The center of the octahedron can also be chosen as an element of the orbit. In this case, the orbit consists of one element and it is transformed according to the identity representation, in which a value 1 is associated with all elements of the group. In many cases, one can choose subspaces in the representation space that are invariant under group transformations. In the case of octahedron, this will be the linear combination $\{s_1 + s_2 + s_3 + s_4 + s_5 + s_6\} /\sqrt{6}$. It's also possible to find a three-dimensional representation. Obviously, under the action of octahedral symmetry operations, the three-dimensional basis $\{s_1 - s_4,\ s_2 - s_5,\ s_3 - s_6\}$ will be transformed as the basis $\{x,\ y,\ z\}$ and the matrices can be easily obtained. Thus, from the basis of the six-dimensional representation, we have managed to obtain two subspaces invariant under group operations: one-dimensional denoted as A_{1g} and three-dimensional denoted as T_{1u}. Also there is a two-dimensional invariant subspace, denoted as E_g. The construction of basis function of E_g symmetry is considered in Chap. 3.

If a group has a subgroup of index 2, then such a subgroup generates a one-dimensional representation in which each element of the subgroup H is assigned the number 1, and each element of the left coset sH is associated with the number -1. To prove it, we must show that the product of the elements of the left coset belongs to the subgroup H. It was shown above that in the case of a subgroup of index 2, the right coset coincides with the left coset. Moreover, since it follows from $s \notin H$ that $s^{-1} \notin H$, we can choose s^{-1} as a representative of the right coset. Then for the elements sh_i and sh_j of the left coset there are such h_k and h_p of H for which the double equality holds:

$$sh_i sh_j = sh_i h_k s^{-1} = h_p \tag{1.2.4}$$

Thus if H is a subgroup of G of index 2, it follows from this equality that the mapping of the group G into the group of one-dimensional matrices which are equal to 1 for $g \in H$ and equal to -1 for $g \in G\backslash H$. (Here $G\backslash H$ denotes the set, called difference of the groups, i.e. it consists of the elements g, such as $g \in G$ and $g \notin H$.)

Example 1.2 Find representations and their characters for the group C_{4v}, i.e. a group of symmetry of a square, whose elements are shown in Fig. 1.1. It was shown above that this group has 3 subgroups of index 2, namely, $\mathsf{C}_4 = \{E, C_{4z}, C_{2z}, C_{4z}^3\}$, $\mathsf{C}_{2v} = \{E, C_{2z}, \sigma_x, \sigma_y\}$, $\mathsf{C}'_{2v} = \{E, C_{2z}, \sigma_a, \sigma_b\}$.Thus, we can build four one-dimensional representations: an identity representation A_1, A_2 representation in which matrices for the subgroup of rotations equal to 1 and others to -1, and representations B_1 and B_2 in which subgroup H consist of elements E, C_{2z} and two reflections in two planes normal to each other.

In order to obtain two-dimensional representation, we consider the result of action of group elements on the vector in three-dimensional space, which can be considered to be the orbit of this group. It is clear that z is invariant under the action of the elements of C_{4v} and transforms as A_1. To find a two-dimensional IR, it is sufficient to consider the action of group elements on the basis $\{x, y\}$. Characters and matrices of irreducible representations of group C_{4v} are presented in Table 1.3. The characters of double-valued or projective representations, which will be considered later on, are also presented in Table 1.3. They are marked with a prime in order to distinguish them from ordinary IRs, defined by formula (1.2.2), which are also called single-valued IRs.

We now turn to the consideration of the general case. If in the n-dimensional space $\{x_1, x_2, ... x_n\}$ of the representation D of the group G, we can single out by a unitary transformation $D'(g) = VD(g)V^{-1}$a k–dimensional ($k < n$) subspace $\{x'_1, x'_2, ... x'_k\}$ invariant under all transformations by group elements, then the representation D is called reducible and is reduced to the form:

Table 1.3 Characters of single-valued and double-valued (denoted by prime) IRs and matrices of IR E of group C_{4v}

	E	C_{2z}	C_{4z}	C_{4z}^3	σ_y	σ_x	σ_a	σ_b
A_1	1	1	1	1	1	1	1	1
A_2	1	1	1	1	-1	-1	-1	-1
B_1	1	1	-1	-1	1	1	-1	-1
B_2	1	1	-1	-1	-1	-1	1	1
E	2	-2	0	0	0	0	0	0
E'_1	2	0	$\sqrt{2}$	$-\sqrt{2}$	0	0	0	0
E'_2	2	0	$-\sqrt{2}$	$\sqrt{2}$	0	0	0	0

$$\underset{E}{\begin{pmatrix}1&0\\0&1\end{pmatrix}}\underset{C_{2z}}{\begin{pmatrix}-1&0\\0&-1\end{pmatrix}}\underset{C_{4z}}{\begin{pmatrix}0&-1\\1&0\end{pmatrix}}\underset{C_{4z}^3}{\begin{pmatrix}0&1\\-1&0\end{pmatrix}}\underset{\sigma_x}{\begin{pmatrix}-1&0\\0&1\end{pmatrix}}\underset{\sigma_y}{\begin{pmatrix}1&0\\0&-1\end{pmatrix}}\underset{\sigma_a}{\begin{pmatrix}0&-1\\-1&0\end{pmatrix}}\underset{\sigma_b}{\begin{pmatrix}0&1\\1&0\end{pmatrix}}$$

$$\begin{pmatrix} D'_{11} & \dots & D'_{1k} & 0 & 0 & 0 \\ \dots & \dots & \dots & 0 & 0 & 0 \\ D'_{k1} & D'_{k2} & D'_{kk} & 0 & 0 & 0 \\ D'_{k+11} & \dots & D'_{k+1k} & D'_{k+1k+1} & \dots & D'_{k+1n} \\ \dots & \dots & \dots & \dots & \dots & \dots \\ D'_{n1} & \dots & D'_{nk} & D'_{nk+1} & \dots & D'_{nn} \end{pmatrix} \tag{1.2.5}$$

Otherwise, the representation is called reducible. It is easy to show that if the representation is unitary, then the orthogonal complement $\{x'_{k+1}, x'_{k+2},\dots x'_n\}$ of the space $\{x'_1, x'_2,\dots x'_k\}$ is invariant under the transformations D'. This means that the matrix D' has a quasi-diagonal form:

$$\begin{pmatrix} D'_{11} & \dots & D'_{1k} & 0 & 0 & 0 \\ \dots & \dots & \dots & 0 & 0 & 0 \\ D'_{k1} & D'_{k2} & D'_{kk} & 0 & 0 & 0 \\ \dots & \dots & \dots & D'_{k+1k+1} & \dots & D'_{k+1n} \\ \dots & \dots & \dots & \dots & \dots & \dots \\ \dots & \dots & \dots & D'_{nk+1} & \dots & D'_{nn} \end{pmatrix} \tag{1.2.6}$$

A representation whose basis can be decomposed into invariant subspaces is called completely reducible. Thus, a reducible representation is completely reducible (Maschke's theorem). Otherwise, the representation is called irreducible. The term irreducible representation is often used in group theory and is denoted as IR.

1.3 Schur Lemmas and Wigner Theorem

The incredible effectiveness of group theory in quantum mechanics is based on two Schur lemmas and Wigner theorem, which are briefly considered in the present section.

The first Schur lemma. A matrix that commutes with all matrices of an irreducible representation is a multiple of an identity matrix.

Let $D(g)$ be an $n-$dimensional IR of a group G. Assume that there exists a matrix A such that for any g:

$$AD(g) = D(g)A \tag{1.3.1}$$

let $\Phi = \{\varphi_1, ..\varphi_n\}$ be a self-vector of matrix A in the space of representational $D(g)$:

$$A\Phi = \lambda\Phi \tag{1.3.2}$$

Consider the space of vectors obtained from the vector Φ by transformations of the group G:

$$D(g)\Phi = \Phi_g \tag{1.3.3}$$

Acting on the vector Φ_g with the matrix A and taking into account the condition of the lemma, we obtain that this vector is also an eigenvector of $D(g)$:

$$A\Phi_g = AD(g)\Phi = D(g)A\Phi = \lambda\Phi_g \tag{1.3.4}$$

This implies that the space of eigenvectors of the matrix A corresponding to one eigenvalue is invariant under transformations by the matrices $D(g)$. By the condition that the representation $D(g)$ is irreducible, the space of eigenvectors of the matrix A must coincide with the basis of the IR D. Therefore, the matrix A, whose action reduces to multiplication by the same number, must be a multiple of the identity matrix.

The second Schur lemma. Let $D^1$and D^2 be two non-equivalent irreducible representations of a group G of dimensions n_1 and n_2. Then any rectangular matrix A for which the condition is satisfied:

$$AD^1(g) = D^2(g)A \tag{1.3.5}$$

for all $g \in G$ is a zero matrix. Consider the case of unitary representations. Let us take the Hermitian conjugates of both parts of this equality and use the unitarity of the IR:

$$\left(D^1(g)\right)^{-1} A^+ = A^+ \left(D^2(g)\right)^{-1} \tag{1.3.6}$$

From the group multiplication law it follows that the inverse matrix corresponds to the inverse element, so we have:

$$D^1(g^{-1})A^+ = A^+D^2(g^{-1}) \tag{1.3.7}$$

Since g runs through the entire group, the last equality implies that:

$$D^1(g)A^+ = A^+D^2(g) \tag{1.3.8}$$

Multiplying both sides of the equation on the left by A and transforming the left side of the equation using the assumption (1.3.5), we obtain

$$D^2(g)AA^+ = AA^+D^2(g) \tag{1.3.9}$$

Then, according to the first Schur lemma, we have

$$AA^+ = E_{n_2} \tag{1.3.10}$$

Let's consider two cases.

1. The dimensions n_1 and n_2 coincide, but the representations are not equivalent by the condition. In this case, from the conditions of the lemma we obtain that

$$D^1(g) = A^{-1}D^2(g)A \tag{1.3.11}$$

This means that the representations are equivalent, which contradicts the condition of the lemma. Hence, the matrix A must be a zero matrix.

2. The dimensions n_1 and n_2 are different. Let us suppose that $n_2 > n_1$. In this case the matrix A is rectangular. Let's complement the matrix A with $n_2 - n_1$ zero columns, and the matrix A^+ with $n_2 - n_1$ with zero rows. These extended square matrices will be denoted by a wave above. It follows from the construction that the matrix $\tilde{A}\tilde{A}^+$, as well as the matrix AA^+, satisfies the relation similar to (1.3.9):

$$D^2(g)\tilde{A}\tilde{A}^+ = \tilde{A}\tilde{A}^+D^2(g) \tag{1.3.12}$$

From this relation and the first Schur lemma, it follows that for new matrices, the same equality holds as for the original rectangular matrices:

$$\tilde{A}\tilde{A}^+ = E_{n_2} \tag{1.3.13}$$

However, according to the construction, $det(\tilde{A}) = 0$, so in order to hold the relation (1.3.12) matrix A must be a zero matrix.

Wigner theorem. This is the main theorem connecting group theory and quantum mechanics which justifies *unreasonable effectiveness* [55] of group theory in physics and in the theory of electron structure. Consider the transformation of the eigenfunctions of the Hamiltonian $\hat{H}$ under the action of elements of group G. The symmetry of a quantum mechanical system means that the Hamiltonian is invariant with respect to the transformation group G. Let $\Phi_\epsilon = \{\phi_i, i = 1, ..n\}$ be the wavefunctions corresponding to the eigenvalue of the Hamiltonian ϵ. The action of a group element on the wavefunction results in a transformation of the basis set:

$$g\Phi_\epsilon(x) = D(g)\Phi_\epsilon(x) = \Phi'_\epsilon(x) \tag{1.3.14}$$

where $D(g)$ is the IR to which this basis Φ_ϵ belongs. The invariance of Hamiltonian, and hence the Schrödinger equation with respect to the group elements, is written as

$$\hat{H}(gx) = \hat{H}(x) \tag{1.3.15}$$

Consider a Hamiltonian matrix H_{ik} defined as

$$H_{ik} = \int \phi_i^* \hat{H} \phi_k dx \tag{1.3.16}$$

Let us show that the condition for the invariance of the Hamiltonian with respect to the group G is equivalent to the condition for the commutativity of the matrices $D(g)$ and H:

$$HD(g) = D(g)H \tag{1.3.17}$$

When acting with the Hamiltonian operator on the transformed basis and using the previous relation, we get

$$\hat{H}\Phi'_{\epsilon} = HD(g)\Phi_{\epsilon} = D(g)H\Phi_{\epsilon} = D(g)\epsilon\Phi_{\epsilon} = \epsilon\Phi'_{\epsilon} \tag{1.3.18}$$

Let $\Gamma(g)$ be some reducible representation of the group G. Using a unitary transformation, we reduce it to a quasi-diagonal form:

$$\Gamma(g) = \begin{pmatrix} D^1(g) & 0 & 0 \\ 0 & D^2(g) & 0 \\ 0 & 0 & D^3(g) \end{pmatrix} \tag{1.3.19}$$

There can also be repeated IRs on the main diagonal. Let some matrix H commute with all matrices $\Gamma(g)$:

$$\Gamma(g)H = H\Gamma(g) \tag{1.3.20}$$

The matrices $\Gamma(g)$ and H have a block structure. A block at the intersection of an i-block row and a k-block column has n_i rows and n_k columns. Blocks of the matrix H, denote $H_{\mu\nu}$. Taking into account the block-diagonal structure of the matrix $D(g)$, we write the commutation condition in the form:

$$D^{\mu}(g)H_{\mu\nu} = D^{\nu}(g)H_{\nu\mu} \tag{1.3.21}$$

It follows from the first Schur lemma that for $\mu = \nu$ the matrices $H_{\mu\mu}$ are multiples of the identity matrices. It follows from the second Schur lemma that for $\mu \neq \nu$ the matrices $H_{\mu\nu}$ are zero matrices. It also follows that in the absence of accidental degeneracy, each IR has its own energy value (a corollary from the first Schur lemma) and that the states belonging to different IRs do not interact (a corollary from the second Schur lemma).

It follows from the proved theorem that the fundamental problem of quantum mechanics is to find the basis functions of irreducible representations. Let the system have symmetry G and consist of two weakly interacting subsystems α and β, and the wavefunctions of each of the subsystems are known. Let φ^{ν}_{α} be the wavefunctions of the first subsystem that belong to the IR D^{ν}, and φ^{ν}_{β} be the wavefunctions of the second subsystem that belong to the same IR. Then all wavefunctions ψ^{ν}_{α} correspond to one eigenvalue ϵ^{ν}_{α} of the Hamiltonian $\hat{H}_{\alpha}$ of the first subsystem, and all wavefunctions ψ^{ν}_{β} correspond to the eigenvalue ϵ^{ν}_{β} of the Hamiltonian $\hat{H}_{\beta}$ of the second subsystem. In this case, it is required to find additional quantum numbers that make it possible to physically separate the bases belonging to one IR. Let us consider the interaction between subsystems α and β as a perturbation with perturbation operator $\hat{H}_{\alpha\beta}$. It follows from the Wigner theorem that the interaction between the states of two systems, φ^{ν}_{α} and φ^{μ}_{β}, belonging to different IRs D^{ν} and D^{μ}, is equal to zero. Therefore, the problem of finding the eigenfunctions of the whole Hamiltonian $\hat{H}_{\alpha} + \hat{H}_{\beta} + \hat{H}_{\alpha\beta}$ is solved separately for each one of the IRs of symmetry group.

Consider the case when each IR D^{ν} appears only once, i.e. the case without accidental degeneracy. Since all eigenfunctions φ_{ν} of Hamiltonian H_0 have the same energy ϵ^{ν}, these eigenvalues don't depend on unitary transformation. However, the energy may depend on a unitary transformation due to additional perturbation Hamiltonian H'. In Chap. 4, additional quantum numbers for two-dimensional IRs of Cooper pairs will be considered. These quantum numbers define unitary transformation of IR and different structures of basis sets belonging to one IR. These additional quantum numbers are connected with the energy due to the additional small interaction H'. Namely, the Hamiltonian H_0 of a Cooper pair corresponds to the interaction between states k and $-k$, connected by space inversion, but additional quantum numbers are essential for a perturbation Hamiltonian H', corresponding to small interaction of the states with almost parallel $k-$ vectors, connected by reflection in a symmetry plane.

1.4 Orthogonality Relations for Matrix Elements of Representations

Let D^{α} and D^{β} be two non-equivalent unitary IRs of a group G of dimensions n_{α} and n_{β}. Let us prove that the following two relations exist between the matrix elements of IRs, which are called the orthogonality relations:

$$\sum_{g\in G} D^{\alpha}_{ij}(g) D^{\beta *}_{kl}(g) = 0 \tag{1.4.1}$$

$$\sum_{g\in G} D^{\alpha}_{ij}(g) D^{\alpha *}_{kl}(g) = \frac{|G|}{n_{\alpha}} \delta(i,k)\delta(j,l) \tag{1.4.2}$$

To prove orthogonality relation (1.4.1), we define the matrix A as follows:

$$A_{ik} = \sum_{g\in G} D^{\alpha}_{ip}(g) X_{pq} D^{\beta}_{qk}(g^{-1}) \tag{1.4.3}$$

where X is an arbitrary rectangular matrix with n_{α} rows and n_{β} columns. Let us find the rules for commutation of the matrix A with the IRs of the group G:

$$D^{\alpha}(\tilde{g})A = D^{\alpha}(\tilde{g}) \sum_{g\in G} D^{\alpha}(g) X D^{\beta}(g^{-1}) = \sum_{g\in G} D^{\alpha}(\tilde{g}) D^{\alpha}(g) X D^{\beta}(g^{-1}) D^{\beta}(\tilde{g}^{-1}) D^{\beta}(\tilde{g}) =$$

$$= \sum_{g\in G} D^{\alpha}(\tilde{g}g) X D^{\beta}\left(\left(\tilde{g}g\right)^{-1}\right) D^{\beta}(\tilde{g}) = A D^{\beta}(\tilde{g}) \tag{1.4.4}$$

In the proof, we used the definition of the matrix A and the fact that if g runs through the whole group, then the product $\tilde{g}g$ also runs through the whole group. Then, according to the second Schur lemma, the matrix A is a zero matrix, i.e.

$$\sum_{g\in G}\sum_{pq} D^{\alpha}_{ip}(g)X_{pq}D^{\beta}_{qk}(g^{-1}) = 0 \tag{1.4.5}$$

By condition, X is an arbitrary matrix, so we assume that this matrix has only one element equal to one:

$$X_{pq} = \delta(p, j)\delta(q, l) \tag{1.4.6}$$

Then, we have

$$\sum_{g\in G} D^{\alpha}_{ij}(g)D^{\beta}_{lk}(g^{-1}) = 0 \tag{1.4.7}$$

From the resulting formula, taking into account the unitarity of the representations, we obtain the first of the orthogonality relations:

$$\sum_{g\in G} D^{\alpha}_{ij}(g)D^{\beta *}_{kl}(g) = 0 \tag{1.4.8}$$

The proof of the second of the orthogonality relation (1.4.2) is similar. In this case, in the definition of the matrix A, it is necessary to take the same IR, for example, D^{α}. In a similar way, we obtain that the matrix A commutes with all the matrices of IR D^{α}, therefore it is a multiple of the identity matrix:

$$\sum_{g\in G}\sum_{pq} D^{\alpha}_{ip}(g)X_{pq}D^{\alpha}_{qk}(g^{-1}) = \lambda\delta(i, k) \tag{1.4.9}$$

We now define the matrix X_{pq} by the relation $X_{pq} = \delta\,(p, j)\,\delta(q, l)$. Then, introducing the notation λ_{jl}, we obtain

$$\sum_{g\in G} D^{\alpha}_{ij}(g)D^{\alpha}_{lk}(g^{-1}) = \lambda_{jl}\delta(i, k) \tag{1.4.10}$$

Next, we set $i = k$ and sum the left side over k:

$$\sum_{g\in G}\sum_{k} D^{\alpha}_{kj}(g)D^{\alpha}_{lk}(g^{-1}) = \sum_{g\in G}\sum_{k} D^{\alpha}_{lk}(g^{-1})D^{\alpha}_{kj}(g) = \sum_{g\in G} D^{\alpha}_{lj}(E) = |G|\,\delta(j, l) \tag{1.4.11}$$

Summing up the right side for $i = k$ we get $\lambda_{jl} n_\alpha$. Equating both parts, we get

$$\lambda_{jl} = \delta(j, l) \frac{|G|}{n_\alpha} \tag{1.4.12}$$

Substituting this relation into (1.4.10) and making use unitarity, i.e. $D^\alpha_{lk}(g^{-1}) = \left(D^\alpha_{lk}(g)\right)^{-1} = D^{\alpha *}_{kl}(g)$, we obtain (1.4.2).

1.5 Classes and Characters

For a certain element r and all elements g of the group G, consider the set of conjugate elements grg^{-1}. When g runs through the entire group, the conjugate elements of r run through a set of dimension less than $|G|$, called the class of conjugate elements. It is obvious that the identity element E forms one class and that each element of an Abelian group also forms a separate class. Thus a group C_4 consisting of four elements $\{E, C_4, C_2, C_4^3\}$ consists of four classes.

If an element v is conjugated to element r, and w is conjugated to element v:

$$v = g_1 r g_1^{-1},\ w = g_2 v g_2^{-1},$$

then

$$w = g_2 g_1 r g_1^{-1} g_2^{-1} = g_2 g_1 r \left(g_2 g_1\right)^{-1},$$

i.e. element w is conjugated to element r.

Classes of a group are non-intersecting sets, whose total dimension equals to the number of elements in group $|G|$. For point symmetry groups, conjugate elements have a simple physical meaning.

Consider for example classes of group of pure rotation of cube O, which has axes of second, third and fourth orders, shown in Fig. 1.7. Rotations through the same angles around different axes belong to the same class only if there are elements of the group that translate these axes into each other. For example, in the symmetry group of the cube O, six rotations C_4 and C_4^3 around three fourfold axes belong to one class. Three rotations C_2 around fourfold axis form a separate class, since in group O there are no elements rotating fourfold to positions of twofold axes. Six rotation C_2 around axis passing through the midpoints of opposite edges belong to another class. Also eight rotation C_3 and C_3^2 form a class. Thus, total number of elements in group O equals to 24. Adding space inversion I doubles number of group elements and classes. Inversion I forms a separate class. Multiplication of rotations C_2 by I result in reflection σ_h in plane normal to this axis. For rotation C_3 (or C_4) we write $IC_3 = IC_2C_2C_3 = \sigma_h C_2 C_3$. Such element is called mirror reflection and is denoted S_3. Thus, we obtain 10 classes of group O_h presented in Table 1.4.

If we subject all matrices of the representation $D(g)$ of the group to the unitary transformation $VD(g)V^{-1}$, then we obtain a representation equivalent to the original

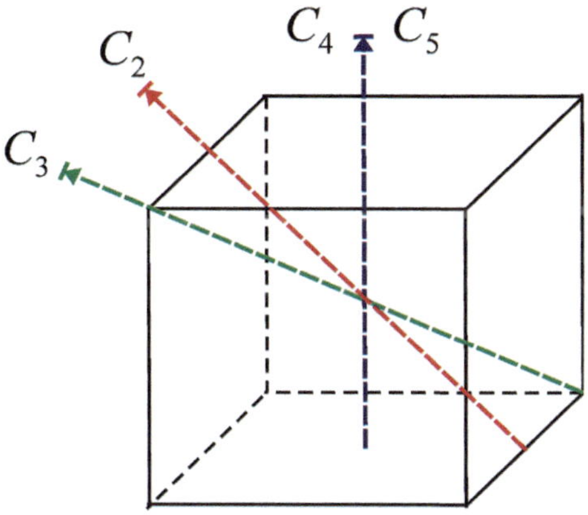

Fig. 1.7 Classes of group O

Table 1.4 Classes of group O_h

C_1	C_2	C_3	C_4	C_5	C_6	C_7	C_8	C_9	C_{10}
E	$8C_3$	$6C_2$	$6C_4$	$3C_2\ (= C_4^2)$	I	$8S_6$	$6\sigma_d$	$6S_4$	$3\sigma_h$

one, i.e. with the same group law of multiplication. With this transformation, the trace of the matrix (the sum of the diagonal elements) is preserved:

$$\sum_i \left(VD(g)V^{-1}\right)_{ii} = \sum_{ikl} V_{ik}D_{kl}(g)V_{li}^{-1} = \sum_{kl} \delta_{kl}D_{kl}(g) = \sum_k D_{kk}(g) \quad (1.5.1)$$

In group theory, the trace of a representation matrix $D(g)$ is called the character of the representation, and denoted as $\chi[D(g)]$. If the representation has an index, for example k, then one also writes $\chi^k(g)$.

For a given element g_0, consider the IR matrices D for the class of conjugate elements:

$$D(gg_0g^{-1}) = D(g)D(g_0)D(g^{-1}) = D(g)D(g_0)D^{-1}(g) \quad (1.5.2)$$

It follows from the above consideration and from the unitarity of matrices $D(g)$ that the characters of matrices $D(g)D(g_0)D(g^{-1})$ for all $g \in G$ coincide with the character of the matrix $D(g_0)$. This implies the property of characters: the characters of each of the IRs are the same for all elements of the same class.

Groups of pure rotations by angles divisible by $2\pi/n$ are Abelian (commutative) and are denoted by C_n. All IRs of such groups are one-dimensional. The characters of such groups can be easily found from the condition $(C_n)^n = E$. Table 1.5 shows IRs of group C_4.

Table 1.5 IRs of group C_4

C_4	E	C_4	C_2	C_4^3
A	1	1	1	1
B	1	-1	1	-1
E'	1	i	-1	$-i$
E''	1	$-i$	-1	i

Let us write the orthogonality relation (1.4.2) for the diagonal elements of IRs:

$$\sum_{g \in G} D_{ii}^{\alpha}(g) D_{kk}^{\beta *}(g) = \delta(i, k)\delta(\alpha, \beta)\frac{|G|}{n_{\alpha}} \tag{1.5.3}$$

Summing over i and k, we get

$$\sum_{g \in G} \chi^{\alpha}(g)\chi^{\beta *}(g) = \delta(\alpha, \beta)\,|G| \tag{1.5.4}$$

Since the characters of the elements of one class are the same, the reference books usually present the characters for the classes of the group. In this case, the orthogonality relations are written in terms of the sum over the classes κ of the group:

$$\sum_{\kappa=1}^{K} \chi_{\kappa}^{\alpha}\chi_{\kappa}^{\beta *} n_{\kappa} = \delta(\alpha, \beta)\,|G| \tag{1.5.5}$$

where n_{κ} is a number of elements in a class κ and K is a number of classes.

When denoting $\tilde{\chi}_{\kappa}^{\alpha} = \chi_{\kappa}^{\alpha}\sqrt{n_{\kappa}}$, we give this relation the form of the vector orthogonality relation in the $K-$dimensional space:

$$\sum_{\kappa=1}^{K} \tilde{\chi}_{\kappa}^{\alpha}\tilde{\chi}_{\kappa}^{\beta *} = \delta(\alpha, \beta)\,|G| \tag{1.5.6}$$

Denote by P the number of non-equivalent IRs of the group. The orthogonality relation written above can be satisfied only if $P \leq K$. It is proved in group theory that the number of non-equivalent IRs of a group is equal to the number of its classes.

$$P = K \tag{1.5.7}$$

The characters of the IRs are called simple characters, and the characters of reducible representations are called composite characters. The reducible representation D decomposes into the sum of the IRs D^{α}, standing on the main diagonal of the matrix:

$$D(g) = \sum_{\alpha} f_{\alpha} D^{\alpha}(g) \tag{1.5.8}$$

where factors f_{α} are called frequencies of appearance of IRs D^{α} in the decomposition of a reducible representation D. Thus the character of D is written as

Table 1.6 Characters of IRs of group C_{3v}, as normalized orthogonal three-dimensional vectors

C_{3v}	E	C_3	σ_v
$\tilde{\chi}(A_1)$	1	$\sqrt{2}$	$\sqrt{3}$
$\tilde{\chi}(A_2)$	1	$\sqrt{2}$	$-\sqrt{3}$
$\tilde{\chi}(E)$	2	$-\sqrt{2}$	0

Table 1.7 Characters of IRs of group C_{3v} in a standard form

C_{3v}	E	$2C_3$	$3\sigma_v$
$\chi(A_1)$	1	1	1
$\chi(A_2)$	1	1	-1
$\chi(E)$	2	-1	0

$$\chi\left[D(g)\right] = \sum_{\alpha} f_\alpha \chi^\alpha(g) \tag{1.5.9}$$

Using the orthogonality relations, we obtain a formula for obtaining coefficients f_α, which are also called reduction coefficients:

$$f_\alpha = \frac{1}{|G|} \sum_{g \in G} \chi\left[D(g)\right] \chi^{\alpha *}(g) \tag{1.5.10}$$

As an example, consider the group C_{3v}, i.e. the symmetry group of a regular triangle, which consists of six elements and three classes: the identity element E, a class of rotations C_3 and C_3^2, and the a class consisting of three reflections σ_v in vertical planes. Hence, it follows that the group has three IRs. Pure rotations together with the identity element form an invariant subgroup of index 2, for which we take the identity representation. For the left coset consisting of three elements σ_v, we can take characters equal to +1 and -1. In the case of an invariant subgroup of index 2, the product of elements of the left coset always gives an element of the subgroup. Therefore, the characters chosen in this way correspond to the group multiplication law. These characters, normalized by the number of elements in the class, are given in the first two rows of Table 1.6. Since the sum of squares of dimensions of inequivalent IRs equals to the number of group elements (see next section), we obtain that third IR is two-dimensional and character of identity element equals to 2. It is immediately verified that if we choose the value $\sqrt{2}$ for the normalized character of the second class, the orthogonality relations for all three vectors are fulfilled. Characters of group C_{3v} in standard form are presented in Table 1.7.

1.6 Regular Representation

Let us define the elements R_{ij} of a square matrix of dimension $|G|$ by the formula:

$$R_{ij}(g) = \begin{cases} 1, if\ s_i^{-1} g s_j = E \\ 0, if s_i^{-1} g s_j \neq E \end{cases} \quad (1.6.1)$$

It is easy to verify that the matrices defined by this formula form a group representation and that there is only one element in each row and each column. Let us define the meaning of this representation. The action of an arbitrary element of the group g on an element s_j gives some element $s_i = gs_j$. This operation, called a left shift by element g, rearranges the elements of a group. It follows from the first line of (1.6.1) that $s_i = gs_j$. This means that the representation $R_{ij}(g)$, called regular, implements a permutation of elements of the group isomorphic to the left shift by element g and that $R_{ij}(g)$ is a group representation. It is easy to show that for all $g \neq E$ there are zeros on the main diagonal, while for $g = E$ there are values 1 on the main diagonal. Decomposing a regular representation into irreducible representations, we obtain that each IR occurs in it as many times as its dimension. Since the dimension of the regular representation is equal to the number of elements of the group, we obtain that the sum of squared dimensions of non-equivalent IRs is equal to the number of elements of the group.

$$\sum_k \left|D^k\right|^2 = |G| \quad (1.6.2)$$

where sum runs over all inequivalent IRs of group G and modulus sign || denotes dimension of IR or number of elements in group.

Regular representation is a special case of induced representations, considered in Chap. 3. In the notations of induced representation method, regular representation (1.6.1) is written as $A_{(E)} \uparrow G$, i.e. the identity representation A of subgroup consisting of identity element E, induced in the whole group G.

1.7 Atomic Wavefunctions in Point Group Symmetry

The Hamiltonian of an atom is invariant not only with respect to the group of pure rotations $\mathsf{O}^+(3)$, but also with respect to the group $\mathsf{O}(3)$, i.e. $\mathsf{O}^+(3)$ group extended by the inversion I:

$$\mathsf{O}(3) = \mathsf{O}^+(3) \times \mathsf{C}_i \quad (1.7.1)$$

where $\mathsf{C}_i = \{E, I\}$. Since the inversion commutes with all rotations of $\mathsf{O}^+(3)$, the IRs of the whole $\mathsf{O}(3)$ group are obtained by multiplying the IRs of the $\mathsf{O}^+(3)$, designated as D^l, by the IRs of the group $\mathsf{C}_i = \{E, I\}$, what reduces to multiplying the matrices for the elements of the left coset $I\mathsf{O}^+(3)$ by $+1$ or -1. These IRs are called even and odd and are denoted by D^{l+} and D^{l-}, respectively.

The group of three-dimensional rotations is defined as a group of operations that transform the set of points lying on the sphere into itself. It is known that

Table 1.8 Spherical functions Y_l^m ($l = 1$, 2 and 3). The phases are chosen according to Condon and Shortley (see [9])

Y_1^1	$-\sqrt{\frac{3}{8\pi}}\sin\theta e^{i\varphi}$	Y_3^3	$-\sqrt{\frac{35}{64\pi}}\sin^3\theta e^{3i\varphi}$
Y_1^0	$\sqrt{\frac{3}{4\pi}}\cos\theta$	Y_3^2	$\sqrt{\frac{105}{32\pi}}\cos\theta\sin^2\theta e^{2i\varphi}$
Y_1^{-1}	$\sqrt{\frac{3}{8\pi}}\sin\theta e^{-i\varphi}$	Y_3^1	$-\sqrt{\frac{21}{64\pi}}\sin\theta\left(5\cos^2\theta-1\right)e^{i\varphi}$
Y_2^2	$\sqrt{\frac{15}{32\pi}}\sin^2\theta e^{2i\varphi}$	Y_3^0	$\sqrt{\frac{7}{16\pi}}\cos\theta\left(5\cos^2\theta-3\right)$
Y_2^1	$-\sqrt{\frac{15}{8\pi}}\sin\theta\cos\theta e^{i\varphi}$	Y_3^{-1}	$\sqrt{\frac{21}{64\pi}}\sin\theta\left(5\cos^2\theta-1\right)e^{-i\varphi}$
Y_2^0	$\sqrt{\frac{5}{16\pi}}\left(3\cos^2\theta-1\right)$	Y_3^{-2}	$\sqrt{\frac{105}{32\pi}}\cos\theta\sin^2\theta e^{-2i\varphi}$
Y_2^{-1}	$\sqrt{\frac{15}{8\pi}}\sin\theta\cos\theta e^{-i\varphi}$	Y_3^{-3}	$\sqrt{\frac{35}{64\pi}}\sin^3\theta e^{-3i\varphi}$
Y_2^{-2}	$\sqrt{\frac{15}{32\pi}}\sin^2\theta e^{-2i\varphi}$		

under rotations determined by the Euler angles α, β and γ, the spherical harmonics $Y_{lm}(\theta,\varphi)$ are transformed through functions with the same l and all $m = -l, \ldots l$. This means that $2l + 1$ spherical functions form a basis of the IR D^l of the three-dimensional rotation group. The symmetry of spherical functions with respect to the inversion is written as

$$IY_{lm}(\theta,\varphi) = Y_{lm}(\pi-\theta,\varphi+\pi) = (-1)^l Y_{lm}(\theta,\varphi) \tag{1.7.2}$$

This relation may be immediately verified making use of explicit formulas for spherical functions presented in Table 1.8. Hence, it follows that under the action of the elements of O(3) group spherical functions with even l are transformed by even IRs D^{l+}, and spherical functions with odd l are transformed by odd IRs D^{l-}.

The spatial parity (i.e. property under the action of I) of the wave function of atomic term $l^N(^{2S+1}L)$ is $(-1)^{lN}$. From this, it follows that the orbital parts of the wavefunctions of terms of any number of electrons with even angular momentum l are transformed by even IRs D^{L+} of rotation group. If electron angular momentum value l is odd, then the terms of odd number of electrons are odd and terms of even number of electrons are even. For example, the orbital parts of all three terms of p^2 configurations ^1S, ^1D and ^3P are even and all terms of p^3 configuration ^2P, ^2D and ^4S are odd.

If an atom is placed in a crystal field and the symmetry is reduced to a point group, then the IRs of rotation group, in the majority of cases, become reducible and are decomposed into IRs of point groups. If a point group contains an inversion, then its IRs are divided into even and odd relative to inversion, which are denoted by the subscripts g (gerade) or u (ungerade), respectively. Therefore, in crystals with inversion, one-electron wavefunctions with even l are transformed according to the

g-type IRs, and one-electron wavefunctions with odd l are transformed according to the u-type IRs. All crystallographic point groups are subgroups of the O_h and D_{6h} groups, which include inversion. The icosahedron group I_h also contains the inversion. Therefore, to determine the transformation properties of atomic wavefunctions in crystals of symmetry O_h, D_{6h} and I_h, it suffices to consider pure rotations and use subscript g for $s-$ and $d-$functions and subscript u for $p-$ and $f-$functions.

In the group $\mathsf{O}^+(3)$, rotations through the same angle α belong to the same class, so it is sufficient to consider rotation about the z-axis through an angle α. In order to find the characters of the IR $D^l\ (\alpha)$, we use the well-known property of spherical harmonics:

$$Y_l^m(\theta, \varphi + \alpha) = e^{im\alpha} Y_l^m(\theta, \varphi) \tag{1.7.3}$$

The trace of the transformation matrix (character of IR) can be easily estimated:

$$\chi^l(\alpha) = e^{il\alpha} + e^{i(l-1)\alpha} + \ldots + e^{-il\alpha} = e^{-il\alpha} \sum_{n=0}^{2l} \left(e^{i\alpha}\right)^n \tag{1.7.4}$$

$$= \frac{e^{-il\alpha}\left(e^{i\alpha(2l+1)} - 1\right)}{\left(e^{i\alpha} - 1\right)} = \frac{\sin\frac{(2l+1)\alpha}{2}}{\sin\frac{\alpha}{2}} \tag{1.7.5}$$

Note that the transformation matrix of spherical functions will have a diagonal form only when rotating about the $z-$axis. When rotating about other axes, the matrix will change, but the trace of the matrix (character of IR) will remain unchanged.

As an example, consider the case of cubic symmetry $O_h = I \times O$. The group of pure cubic rotations is a subgroup of the group of three-dimensional rotations and contains five classes presented in Table 1.4. Thus characters in l.h.s of (1.7.4) can be easily calculated, and making use of formula (1.5.10) one easily obtains the branching rules for reduction $\mathsf{O}(3) \downarrow O_h$. To obtain complete labels of IRs of O_h group, the subscript g should be added for even $l = 0, 2$ and subscript u for odd $l = 1, 3$. The reduction schemes for point groups O_h, D_{6h} and I_h are presented in Tables 1.9, 1.10 and 1.11, respectively. It is seen from Tables 9 and 11 that all three $p-$ functions form irreducible basis sets of IRs in O_h and I_h symmetry both labeled T_{1u}. In crystal field of O_h symmetry d- functions split on two basis sets E_g and T_{2g}, but in I_h symmetry five $d-$functions form irreducible basis set transforming by IR H_g.

With further symmetry reduction multidimensional irreducible representations can become reducible. It is easy to find a correspondence between IRs with symmetry reduction using the orthogonality relations for point groups. The relations for IRs when reducing O_h to some of its subgroups are given in Table 1.12.

Complex form of spherical functions with magnetic quantum number m presented in Table 1.8 is used in atomic physics. However, in quantum chemistry and condensed matter physics in the majority of cases real form of IRs with real basis sets is used.

Taking into account that the projections on coordinate axes of a unit vector $\boldsymbol{r}$ are $r_x = \sin\theta\cos\varphi$, $r_y = \sin\theta\sin\varphi$, and $r_z = \cos\theta$, and omitting normalization factors we can write for p-functions:

Table 1.9 Characters of transformation matrices of spherical functions l=0, 1, 2, 3 for classes of O_h group and their decomposition into irreducible characters of O_h group and its subgroups T_d and D_{4h}

l	$\chi(D^l)$					IRs of O_h	IRs of T_d	IRs of D_{4h}
	E	$8C_3$	$6C_4$	$6C_2$	$3C_2$			
0	1	1	1	1	1	A_{1g}	A_1	A_{1g}
1	3	0	–1	1	–1	T_{1u}	T_2	$E_u + A_{2u}$
2	5	–1	1	–1	1	$T_{2g} + E_g$	$T_2 + E$	$E_g + B_{2g} + A_{1g} + B_{1g}$
3	7	1	–1	–1	–1	$A_{2u} + T_{1u} + T_{2u}$	$A_1 + T_1 + T_2$	$B_{1u} + E_u + A_{2u} + E_u + B_{2u}$

Table 1.10 Characters of transformation matrices of spherical functions l=0, 1, 2, 3 for classes of D_{6h} group and their decomposition into irreducible characters of D_{6h} group

l	$\chi(D^l)$						IRs of D_{6h}
	C_1	$2C_3$	$2C_6$	C_{2z}	$3C_2$	$3C_2$	
0	1	1	1	1	1	1	A_{1g}
1	3	0	2	–1	–1	–1	$A_{2u} + E_{1u}$
2	5	–1	1	1	1	1	$A_{1g} + E_{1g} + E_{2g}$
3	7	1	–1	–1	–1	–1	$A_{2u} + B_{1u} + B_{2u} + E_{1u} + E_{2u}$

Table 1.11 Characters of transformation matrices of spherical functions l=0, 1, 2, 3 for classes of I_h group and their decomposition into irreducible characters of I_h group

l	$\chi(D^l)$					IR I_h
	C_1	$12C_5$	$12C_5^2$	$20C_3$	$15C_2$	
0	1	1	1	1	1	A_g
1	3	$\frac{1}{2}(1+\sqrt{5})$	$\frac{1}{2}(1-\sqrt{5})$	0	–1	T_{1u}
2	5	0	0	–1	1	H_g
3	7	$-\frac{1}{2}(1+\sqrt{5})$	$\frac{1}{2}(\sqrt{5}-1)$	1	–1	$T_{2u} + G_u$

$p_z = z = \cos\theta = Y_1^0$, $p_x = x = \sin\theta\cos\varphi = \frac{1}{2}\left(Y_1^{-1} - Y_1^1\right)$, $p_y = y = \sin\theta \sin\varphi = \frac{i}{2}\left(Y_1^{-1} + Y_1^1\right)$.

Graphs of square modules of p-orbitals in complex and real representations are shown in Fig. 1.8. For real orbitals, the signs of p_x and p_y are also indicated. The graph of $|Y_{11}|^2 \sim |x+iy|^2$ is independent on φ, but graphs of p_x and p_x strongly depend on φ and have nodes. In O_h symmetry basis set $\{x,\ y, z\}$ belongs to T_{1u}. In D_{4h} symmetry $x-$ and y-functions belong to E_u, and z-function to A_{2u}.

Table 1.12 Correspondence between IRs of O_h and IRs of its subgroups

O_h	T_d	T_h	D_{4h}	C_{4v}	D_{3d}	C_{3v}
A_{1g}	A_1	A_g	A_{1g}	A_1	A_{1g}	A_1
A_{2g}	A_2	A_g	B_{1g}	B_1	A_{2g}	A_2
E_g	E	$E_{1g}+E_{2g}$	$A_{1g}+B_{1g}$	A_1+B_1	E_g	E
T_{1g}	T_1	T_g	$A_{2g}+E_g$	A_2+E	$A_{2g}+E_g$	A_2+E
T_{2g}	T_2	T_g	$B_{2g}+E_g$	B_2+E	$A_{1g}+E_g$	A_1+E
A_{1u}	A_2	A_u	A_{1u}	A_2	A_{1u}	A_2
A_{2u}	A_1	A_u	B_{1u}	B_2	A_{2u}	A_1
E_u	E	$E_{1u}+E_{2u}$	$A_{1u}+B_{1u}$	A_2+B_2	E_u	E
T_{1u}	T_2	T_u	$A_{2u}+E_u$	A_1+E	$A_{2u}+E_u$	A_1+E
T_{2u}	T_1	T_u	$B_{2u}+E_u$	B_1+E	$A_{1u}+E_u$	A_2+E

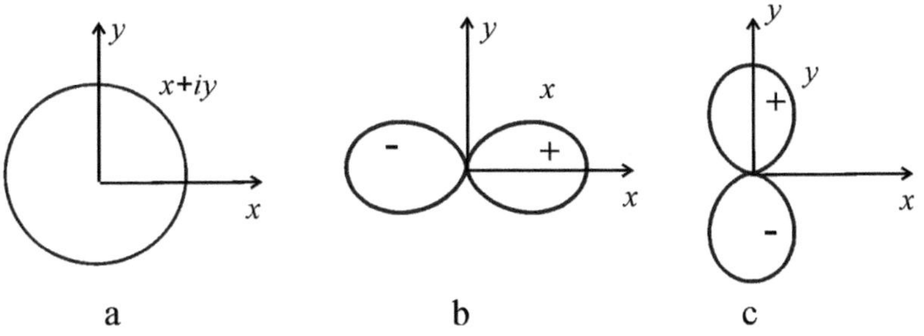

Fig. 1.8 Square modules of $p-$ functions in different representations. The signs of orbitals p_x and p_y are also indicated

In a similar way, one can construct real combinations of atomic d-functions.

$d_{z^2}=Y_2^0=\left(3z^2-1\right), d_{xz}=xz=\frac{1}{2}\left(Y_2^{-1}-Y_2^{-1}\right), d_{yz}=yz=\frac{i}{2}\left(Y_2^1+Y_2^{-1}\right),$
$d_{x^2-y^2}=x^2-y^2=\frac{1}{2}\left(Y_2^2+Y_2^{-2}\right), d_{xy}=xy=\frac{i}{4}\left(Y_2^{-2}-Y_2^2\right)$

Transformation properties of real combinations of atomic $s-$, $p-$ and $d-$functions in some point groups are presented in Table 1.13.

Graphs of orbitals d_{z^2}, $d_{x^2-y^2}$ and d_{xy} are shown in Fig. 1.9. Real combinations $d_{x^2-y^2}$ and d_{xy} have nodes in diagonal and coordinate vertical planes, respectively. The square modules of complex functions $Y_2^{\pm 2}=-\left(x^2-y^2\pm ixy\right)$ are constant for all angles and the graph is similar to that in Fig. 1.8a. Orbital $d_{2z^2-x^2-y^2}$ is even and in D_{4h} symmetry is invariant with respect to rotation about vertical axis and about four axis in basal plane. Thus d_{z^2} belongs to IR A_{1g} of D_{4h}. Orbitals $d_{x^2-y^2}$ and d_{xy} are invariant under the action of I and C_{2z} and change the signs under the action of C_{4z} and C_{4z}^3. The only difference of their symmetry is in reflections in vertical symmetry planes: σ_a and σ_b change the sign of $d_{x^2-y^2}$ and σ_x and σ_y change the sign of d_{xy}. Hence, it follows that in symmetry group D_{4h} $d_{x^2-y^2}$ belongs to B_{1g} and d_{xy}belongs to B_{2g}.

Table 1.13 Atomic $s-$, $p-$ and $d-$functions in O_h, T_d and D_{4h} symmetry

Notation	Function	O_h	T_d	D_{4h}
s		A_{1g}	A_1	A_{1g}
p_z	z	T_{1u}	T_2	A_{2u}
p_y	y	T_{1u}	T_2	E_u
p_x	x	T_{1u}	T_2	E_u
d_{z^2}	$2z^2 - x^2 - y^2$	E_g	E	A_{1g}
$d_{x^2-y^2}$	$x^2 - y^2$	E_g	E	B_{1g}
d_{xy}	xy	T_{2g}	T_2	B_{2g}
d_{xz}	xz	T_{2g}	T_2	E_g
d_{yz}	yz	T_{2g}	T_2	E_g

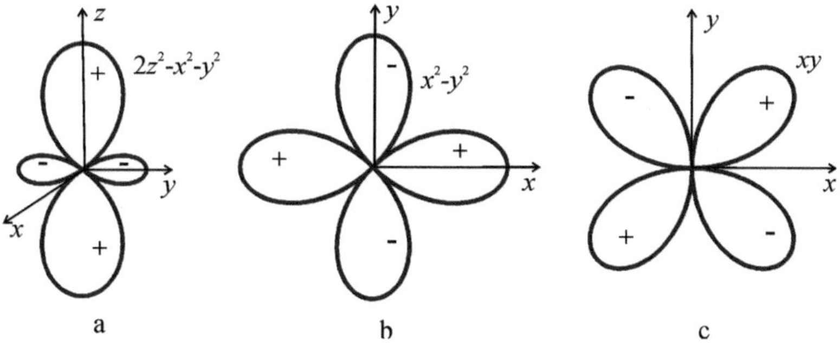

Fig. 1.9 Real $d-$ functions. The remaining two functions d_{xz} and d_{yz} are similar to d_{xy}

Going over to the $f-$functions we see in Table 1.8 that the dependence of functions with $m = 1$ and 2 on angle φ is the same as that of $p-$ and $d-$functions, respectively, but the difference is in the dependence on polar angle θ. The dependence of Y_3^2 in plane normal to $x-$axis is shown in Fig. 1.10a. Two real combinations of Y_3^3 and Y_3^{-3} $x^3 - 3xy^2$ and $y^3 - 3x^2y$ are shown in Fig. 1.10b, c, respectively. Spherical functions in complex and real forms are often used in phenomenological approaches to topological superconductors. Real combinations of atomic $p-$, $d-$ and $f-$functions are used in the calculations of electron structure of molecules and solids. Basis functions constructed from atomic functions for O_h group and some of its subgroups are presented in Tables 1.13 and 1.14.

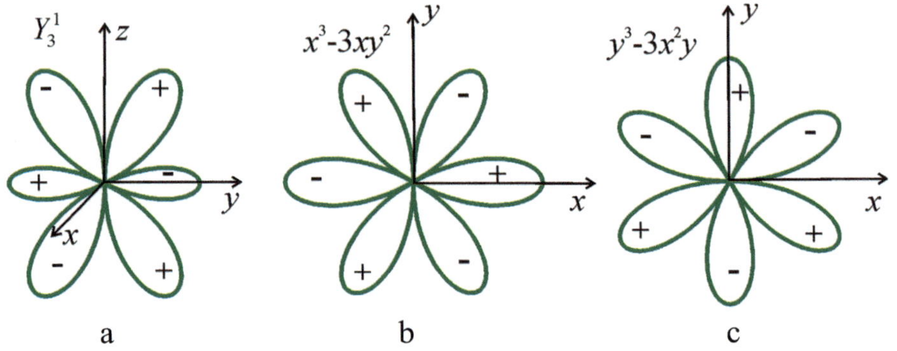

Fig. 1.10 **a** Spherical function Y_3^1 in plane (100), **b** and **c** Two real combinations of Y_3^3 and Y_3^{-3} in plane (001)

Table 1.14 Atomic $f-$functions in O_h, T_d and D_{4h} symmetry

Notation	Function	O_h	D_{4h}	T_d
f_{xyz}	$\sqrt{\frac{105}{16\pi}}xyz$	A_{2u}	B_{1u}	A_1
f_{x^3}	$\sqrt{\frac{7}{16\pi}}z\left(5z^2-3r^2\right)$	T_{1u}	A_{2u}	T_2
f_{y^3}	$\sqrt{\frac{7}{16\pi}}x\left(5x^2-3r^2\right)$	T_{1u}	E_u	T_2
f_{z^3}	$\sqrt{\frac{7}{16\pi}}y\left(5y^2-3r^2\right)$	T_{1u}	E_u	T_2
$f_{x(y^2-z^2)}$	$\sqrt{\frac{105}{16\pi}}x\left(y^2-z^2\right)$	T_{2u}	E_u	T_1
$f_{y(z^2-x^2)}$	$\sqrt{\frac{105}{16\pi}}y\left(z^2-x^2\right)$	T_{2u}	E_u	T_1
$f_{z(x^2-y^2)}$	$\sqrt{\frac{105}{16\pi}}z\left(x^2-y^2\right)$	T_{2u}	B_{2u}	T_1

1.8 Projective Representations and Spinors in Point Group Symmetry

In group theory, each symmetry element g is associated with a representation matrix $D(g)$ such that the effect of the symmetry element on the wavefunction is brought to multiplying the state vector $\{\psi\}$ by the IR matrix $D(g)$. However, it is known from quantum mechanics that the state of a system is described not by the vector $\{\psi\}$, but by the ray $\omega\,\{\psi\}$, where ω is complex phase factor equal in absolute value to one. Therefore, the elements of group can be associated with more general operators, which transform rays into rays. In this case, the well-known rule of multiplication of ordinary representations 1.2.2 is replaced by a more general one:

$$\hat{D}(g_i)\hat{D}(g_j) = \omega_{i,j}\hat{D}(g_k),\ g_i g_j = g_k \tag{1.8.1}$$

Factors $\omega_{i,j}$ that are complex numbers, modulo equal to one, form a factor system of projective (ray) representation $\hat{D}$. Factors cannot be chosen arbitrarily, because the associative law of group multiplication leads to the fact that n^2 factors $\omega_{i,j}$ are connected by n^3 conditions:

$$\omega_{i,j}\omega_{ij,p} = \omega_{i,jp}\omega_{j,p} \tag{1.8.2}$$

Two-factor systems may be projectively equivalent and projectively non-equivalent. Let the representation $\hat{D}(g)$ belong to the factor system $\omega_{i,j}$. Let $\omega'_{i,j}$ be another factor system. If it is possible to find constants c_i for all matrices $\hat{D}(g_i)$ such that the matrices $\hat{D}'(g_i) = c_i\ \hat{D}(g_i)$ form a projective representation with factor system $\omega'_{i,j}$, then such two-factor systems are said to be projectively equivalent. In this case, we have

$$\hat{D}'(g_i)\hat{D}'(g_j) = c_i c_j \hat{D}(g_i)\hat{D}(g_j) = c_i c_j \omega_{i,j}\hat{D}(g_i g_j) = \frac{c_i c_j \omega_{i,j}\hat{D}'(g_i g_j)}{c_{ij}} \tag{1.8.3}$$

Thus, the relationship between these factor systems is given by the formula:

$$\omega'_{i,j} = \frac{c_i c_j \omega_{i,j}}{c_{ij}} \tag{1.8.4}$$

In quantum mechanics in the case of a half-integer spin, the double-valued representations of double groups are usually considered. In this technique, the number of group elements doubles. However, instead of introducing a double group, it is convenient to introduce projective representations of the same group. The factors ω_{ij} of this factor system are equal to -1, if the product of the elements $g_i g_j$ gives the element g_{ij} with an additional rotation by 2π or larger, and the factors equal to $+1$ in all other cases. For example, $C_4 \times C_4 = C_2$ and $\omega = 1$. But $C_4^3 \times C_4^3 = C_2 + 2\pi$ and $\omega = -1$. All such cases are noted in the multiplication tables of group elements given in the book by Kovalev [60]. The concepts of reducibility and irreducibility for projective representations are the same as for ordinary ones. The orthogonality relations for projective representations with the same factor systems are written in the same way as for ordinary IRs. The sum of the squared dimensions of all non-equivalent projective IRs with any factor system, as well as in the case of ordinary IRs, is equal to the number of elements in the group. However, since the division into classes in the case of projective IRs differs from that for point groups, the dimensions of projective IRs themselves differ from the dimensions of ordinary IRs of the same group [57]. Projective representations are also used to construct IRs of non-symmorphic space groups in the cases when the wavevector is on faces of a Brillouin zone (see Chap. 3).

The fundamental difference between the description of the spin part of the wavefunction and its coordinate part is related to the following. The spatial part of the wavefunction is determined by the coordinate $\boldsymbol{r} = (x, y, z)$, and the action of the symmetry operation R on the wavefunction is associated with the transformation of coordinates by the following relationship:

$$R\Psi(\boldsymbol{r}) = \Psi(R^{-1}\boldsymbol{r}) \tag{1.8.5}$$

In non-relativistic approach, total wavefunction $\Phi(\boldsymbol{r}, \bar{s})$ may be represented as a direct product of its spatial part $\Psi(\boldsymbol{r})$ and spin part $\bar{s}$. The spin part called spinor is a two-component complex quantity $\bar{s} = \begin{pmatrix} \chi_1 \\ \chi_2 \end{pmatrix}$. However, strictly speaking, similar relation for total wave function:

$$R\Psi(\boldsymbol{r})\bar{s} = \Psi(R^{-1}\mathbf{r})R^{-1}(\bar{s}) \tag{1.8.6}$$

is not defined.

Albeit the action of the operator R on total wavefunction may be written in a more general form:

$$R\Psi(\boldsymbol{r})\bar{s} = \sum_j D_{ij}^{1/2}(R)\chi_j\Psi(R^{-1}\boldsymbol{r}) \tag{1.8.7}$$

The matrices $D_{ij}^{1/2}(R)$ must satisfy the definition of group representations. However, these representations are projective. Let $\boldsymbol{r} = (x, y, z)$ be a unit normalized vector. Making use of Pauli matrices this vector may be transformed into two-dimensional Hermitian zero-trace matrix:

$$P(\boldsymbol{r}) = (\boldsymbol{\sigma} \cdot \boldsymbol{r}) = \sigma_x x + \sigma_y y + \sigma_z z = \begin{pmatrix} z & x - iy \\ x + iy & -z \end{pmatrix} \tag{1.8.8}$$

where $\det P(\boldsymbol{r}) = -x^2 - y^2 - z^2 = -1$.

Consider the transformation of this matrix with an arbitrary unitary matrix U:

$$Q = UPU^{\dagger} \tag{1.8.9}$$

The resulting matrix, like the original one, can be represented as

$$Q = \sigma_x x' + \sigma_y y' + \sigma_z z' \tag{1.8.10}$$

Since the unitary transformation U does not change the trace and determinant of the matrix, we have

$$x^2 + y^2 + z^2 = x'^2 + y'^2 + z'^2 \tag{1.8.11}$$

This means that the unitary transformation $Q = UPU^\dagger$ induces an orthogonal coordinate transformation $\boldsymbol{r}' = O(U)\boldsymbol{r}$. Let us now transform the matrix Q by a unitary transformation V: $T = VQV^\dagger = VUPU^\dagger V^\dagger = VUP\,(VU)^\dagger$. This unitary transformation also induces some orthogonal coordinate transformation $\boldsymbol{r}'' = O(V)\boldsymbol{r}'$. Obviously, when sequentially performing orthogonal transformations (product of operations), the same group multiplication law is satisfied as for unitary transformation matrices:

$$O(UV) = O(U)O(V) \tag{1.8.12}$$

Also, $O(E_2) = E_3$. This means that two-dimensional unitary matrices form a representation of an orthogonal group $\mathsf{O}(3)$. However, this representation is not exact. Obviously, if ω is an arbitrary complex number modulo equal to one, and if $V = \omega U$, then $VPV^+ = UPU^+$. Therefore, if we restrict ourselves to unimodular matrices, i.e. matrices whose determinants are equal to one, then we obtain the same set of orthogonal matrices. The group of unitary unimodular matrices is called the special unitary group and in the considered case of two-dimensional matrices is denoted by $SU(2)$. As it is easy to see, in this case, there is still some uncertainty: two unitary unimodular matrices U and $-U$ correspond to the same orthogonal matrix. This means that the rotation (orthogonal transformation of the three-dimensional space) given by the Euler angles (α, β, γ) determines the matrix U up to the sign. Based on the expression of the matrix O of three-dimensional rotations in terms of the Euler angle, we can obtain the following expression for the matrix U:

$$U = \pm \begin{pmatrix} e^{-\frac{1}{2}i(\alpha+\gamma)} \cos \frac{1}{2}\beta & -e^{-\frac{1}{2}i(\alpha-\gamma)} \sin \frac{1}{2}\beta \\ e^{\frac{1}{2}i(\alpha-\gamma)} \sin \frac{1}{2}\beta & e^{\frac{1}{2}i(\alpha+\gamma)} \cos \frac{1}{2}\beta \end{pmatrix} \tag{1.8.13}$$

Similar matrices are defined not only for $s = 1/2$, but also for any half-integer j. Double-valued representations of the rotation group will be denoted by $D^j(R)$, i.e. in the same way as $D^l(R)$, for integer l. For double-valued IRs, the following multiplication rule is valid:

$$D^j(R)D^j(S) = \pm D^j(RS) \tag{1.8.14}$$

Ambiguity also arises under the action of inversion:

$$D^j(IR) = \pm D^j(R) \tag{1.8.15}$$

For such IRs, called double-valued ones, following artificial technique is used: the element F is introduced, the effect of which is to multiply the matrix by -1. This element is associated with rotation by 2π. The number of elements in a group obtained in this way is twice the number of elements in the original group. For example, expanding the group D_2, which consists of the identity element and three rotations about coordinate axis with the angle π, results in a double group, consisting of eight elements, whose double-valued IRs are as follows:

$$\overset{E}{\begin{pmatrix} 1 & 0 \\ 0 & 1 \end{pmatrix}} \overset{F}{\begin{pmatrix} -1 & 0 \\ 0 & -1 \end{pmatrix}} \overset{C_{2x}}{\begin{pmatrix} 0 & i \\ i & 0 \end{pmatrix}} \overset{FC_{2x}}{\begin{pmatrix} 0 & -i \\ -i & 0 \end{pmatrix}} \tag{1.8.16}$$

$$\overset{C_{2y}}{\begin{pmatrix} 0 & 1 \\ -1 & 0 \end{pmatrix}} \overset{FC_{2y}}{\begin{pmatrix} 0 & -1 \\ 1 & 0 \end{pmatrix}} \overset{C_{2z}}{\begin{pmatrix} i & 0 \\ 0 & -i \end{pmatrix}} \overset{FC_{2z}}{\begin{pmatrix} -i & 0 \\ 0 & i \end{pmatrix}} \tag{1.8.17}$$

The multiplication rules for this group are as follows:

$$\begin{aligned} C_{2x}C_{2y} &= FC_{2z}, C_{2y}C_{2x} = C_{2z}, \\ C_{2x}C_{2z} &= C_{2y}, C_{2z}C_{2x} = FC_{2y}, \\ C_{2y}C_{2z} &= FC_{2x}, C_{2z}C_{2y} = C_{2x}, \\ C_{2x}^2 &= C_{2y}^2 = C_{2z}^2 = F \end{aligned}$$

It can be seen from the last triple relation that rotations by 2π lead to the element F. Then the rotation by 4π corresponds to the element $F^2 = E$. However, for all calculations, it is sufficient to consider matrices of projective IRs of point group elements and perform summation in orthogonality relation over the point group only, i.e. without its extension with element F. By calculation of products of matrices of double-valued IRs for point group elements in (1.8.17) one easily obtains factor system of the projective representation of D_2.

Notations of double-valued IRs of point groups are similar to that of single-valued IRs, but with a prime, i.e. E', and G' stand for two- and four-dimensional IRs, respectively. In a general case, projective representations are denoted by cover throughout. Due to the relation (1.8.15) double-valued IRs have no indexes g and u. In a strong spin-orbit coupling case, atomic wavefunctions have half-integer angular momentum value j. Under the action of elements of rotation group $\mathsf{O}^+(3)$ the functions with half-integer angular momentum j are transformed by the matrices $\hat{D}^j$, whose characters are written as

$$\chi^j(\alpha) = \frac{\sin\left[\left(j + \frac{1}{2}\right)\alpha\right]}{\sin\frac{\alpha}{2}} \tag{1.8.18}$$

It follows from this formula that for a half-integer j:

$$\chi^j(\alpha) = -\chi^j(\alpha + 2\pi) \tag{1.8.19}$$

This means that rotation by 2π changes the character sign. The values of the characters for angles equal to 0 and 2π are obtained according to L'Hopital's rule:

$$\chi^j(0) = \frac{(j + 1/2)\cos\left[(j + 1/2)\cdot 0\right]}{1/2\cos\left[(1/2)\cdot 0\right]} = 2j + 1 \tag{1.8.20}$$

$$\chi^j(2\pi) = \frac{(j+1/2)\cos\left[(j+1/2)\cdot 2\pi\right]}{1/2\cos\left[(1/2)\cdot 2\pi\right]} = -(2j+1) \tag{1.8.21}$$

The latter relation is valid, since j is half-integer.
The only matching characters are the rotations by π and 3π:

$$\chi^j(\pi) = \chi^j(3\pi) = 0 \tag{1.8.22}$$

Decomposition of IRs $\hat{D}^j$ of rotation group $\mathsf{O}^+(3)$ with half-integer j, corresponding to reduction of symmetry from rotation group $\mathsf{O}^+(3)$ to octahedral and hexagonal groups O and D_6 are presented in Tables 1.15 and 1.16, respectively.

In the case of strong spin-orbit coupling atomic $p-$ functions in rotation symmetry $\mathsf{O}^+(3)$ have total angular moments $j = 1/2$ and $j = 3/2$, and $d-$functions have $j = 3/2$ and $j = 5/2$. The ground state is the state with the smallest total angular momentum. It is seen from Table 1.15 that as in the case of weak spin-orbit coupling, $p-$functions don't split in O_h symmetry: $p_{1/2}$ functions belong to two-dimensional IR E_1' and $p_{3/2}$ functions belong to four-dimensional IR G'. The functions $d_{3/2}$ also don't split in cubic field, but the function $d_{5/2}$ splits into two states. In D_6 symmetry (see Table 1.16) the state $j = 1/2$ doesn't split, but the state $j = 3/2$ splits into two components.

The matrices of $SU(2)$ transform two-dimensional vectors $\bar{s} = \begin{pmatrix} \chi_1 \\ \chi_2 \end{pmatrix}$, called spinors. It is known from quantum mechanics that every spinor corresponds to spin projection P_i on coordinate axes, defined as

Table 1.15 Correspondence between IRs $\hat{D}^j$ of rotation group $\mathsf{O}^+(3)$ with half-integer j and double-valued IRs of octahedral group O

j	IR O
1/2	E_1'
3/2	G'
5/2	$E_2' + G'$
7/2	$E_1' + E_2' + G'$
9/2	$E_1' + 2G'$
11/2	$E_1' + E_2' + G$

Table 1.16 Correspondence between IRs $\hat{D}^j$ of rotation group $\mathsf{O}^+(3)$ with half-integer j and double-valued IRs of group D_6

j	IR D_6
1/2	E_1'
3/2	$E_1' + E_3'$
5/2	$E_1' + E_2' + E_3'$
7/2	$E_1' + 2E_2' + E_3'$
9/2	$E_1' + 2E_2' + 2E_3'$
11/2	$2E_1' + 2E_2' + 2E_3'$

$$P_i = \bar{s}^{\dagger}\sigma_i\bar{s},\ i = x, y, z \quad (1.8.23)$$

where σ_i are Pauli matrices and the sign $\dagger$ denotes Hermitian conjugation, i.e. matrix transposed and complex conjugate.

It is clear from this formula that spinor multiplication by complex phase factor doesn't change a projection. For some selected spinors, we can write

$$P_x = \frac{1}{2}(1, 1)\begin{pmatrix} 0 & 1 \\ 1 & 0 \end{pmatrix}\begin{pmatrix} 1 \\ 1 \end{pmatrix} = 1 \quad (1.8.24)$$

$$P_y = \frac{1}{2}(1, -i)\begin{pmatrix} 0 & -i \\ i & 0 \end{pmatrix}\begin{pmatrix} 1 \\ i \end{pmatrix} = 1 \quad (1.8.25)$$

$$P_z = (1, 0)\begin{pmatrix} 1 & 0 \\ 0 & -1 \end{pmatrix}\begin{pmatrix} 1 \\ 0 \end{pmatrix} = 1 \quad (1.8.26)$$

It follows from the above formulas that the spin states polarized in directions x, y and z are represented by spinors $\bar{s}_x = \frac{1}{\sqrt{2}}\begin{pmatrix} 1 \\ 1 \end{pmatrix}$, $\bar{s}_y = \frac{1}{\sqrt{2}}\begin{pmatrix} 1 \\ i \end{pmatrix}$ and $\bar{s}_z = \begin{pmatrix} 1 \\ 0 \end{pmatrix}$, respectively. It can be immediately verified that the projections of the spin states represented by these spinors on other axes are zero.

Example 1.3 Transformations of spinors by the elements of D_4 group.

Double-valued IRs of D_4 group are presented in Table 1.17.

1) The other double-valued IR E_2' of D_4 is obtained as a direct product $E_2' = E_1' \times B_1$.

The factor system of projective IR E_1' can be easily obtained by multiplication:

$$\hat{D}(C_{4z}^3)\hat{D}(C_{4z}^3) = \tfrac{1}{2}\begin{pmatrix} 1+i & 0 \\ 0 & 1-i \end{pmatrix}\begin{pmatrix} 1+i & 0 \\ 0 & 1-i \end{pmatrix} = \begin{pmatrix} i & 0 \\ 0 & -i \end{pmatrix} = -\hat{D}(C_{2z})$$

$$\hat{D}(C_{4z})\hat{D}(C_{4z}) = \tfrac{1}{2}\begin{pmatrix} -1+i & 0 \\ 0 & -1-i \end{pmatrix}\begin{pmatrix} -1+i & 0 \\ 0 & -1-i \end{pmatrix} = \begin{pmatrix} -i & 0 \\ 0 & i \end{pmatrix} = \hat{D}(C_{2z})$$

Table 1.17 Double-valued IR E_1' and single-valued IR B_1 of group D_4 [1)]

	E	C_{4z}^3	C_{2a}	C_{2z}
E_1'	$\begin{pmatrix} 1 & 0 \\ 0 & 1 \end{pmatrix}$	$\frac{1}{\sqrt{2}}\begin{pmatrix} 1+i & 0 \\ 0 & 1-i \end{pmatrix}$	$\frac{1}{\sqrt{2}}\begin{pmatrix} 0 & -1-i \\ 1-i & 0 \end{pmatrix}$	$\begin{pmatrix} -i & 0 \\ 0 & i \end{pmatrix}$
B_1	1	-1	-1	1
	C_{2x}	C_{4z}	C_{2b}	C_{2y}
E_1'	$\begin{pmatrix} 0 & -i \\ -i & 0 \end{pmatrix}$	$\frac{1}{\sqrt{2}}\begin{pmatrix} -1+i & 0 \\ 0 & -1-i \end{pmatrix}$	$\frac{1}{\sqrt{2}}\begin{pmatrix} 0 & -1+i \\ 1+i & 0 \end{pmatrix}$	$\begin{pmatrix} 0 & -1 \\ 1 & 0 \end{pmatrix}$
B_1	1	-1	-1	1

The action of C_{2a} on spinors, polarized in x- and y-directions, results in spinors, polarized in y- and x-axis, respectively:

$$\frac{1}{\sqrt{2}}\begin{pmatrix} 0 & -1-i \\ 1-i & 0 \end{pmatrix}\frac{1}{\sqrt{2}}\begin{pmatrix} 1 \\ 1 \end{pmatrix} = \frac{1}{2}\begin{pmatrix} -1-i \\ 1-i \end{pmatrix} = \frac{1}{\sqrt{2}}\begin{pmatrix} e^{\frac{i5\pi}{4}} \\ e^{\frac{i7\pi}{4}} \end{pmatrix} = \frac{e^{\frac{i5\pi}{4}}}{\sqrt{2}}\begin{pmatrix} 1 \\ i \end{pmatrix}$$

$$\frac{1}{\sqrt{2}}\begin{pmatrix} 0 & -1-i \\ 1-i & 0 \end{pmatrix}\frac{1}{\sqrt{2}}\begin{pmatrix} 1 \\ i \end{pmatrix} = \frac{1}{2}\begin{pmatrix} 1-i \\ 1-i \end{pmatrix} = \frac{1}{\sqrt{2}}e^{\frac{i7\pi}{4}}\begin{pmatrix} 1 \\ 1 \end{pmatrix}$$

Also for the spinor, polarized in z-direction the action of C_{2a} results in:

$$\frac{1}{\sqrt{2}}\begin{pmatrix} 0 & -1-i \\ 1-i & 0 \end{pmatrix}\begin{pmatrix} 1 \\ 0 \end{pmatrix} = \frac{1}{\sqrt{2}}\begin{pmatrix} 0 \\ 1-i \end{pmatrix} = e^{\frac{i7\pi}{4}}\begin{pmatrix} 0 \\ 1 \end{pmatrix}$$

what corresponds the change of polarization from z to $-z$.

For comparison, the transformation of a three-dimensional vector $\boldsymbol{r} = (x, y, z)$ with the element C_{2a} of group D_4 is written as

$$E(C_{2a})\begin{pmatrix} x \\ y \end{pmatrix} = \begin{pmatrix} 0 & 1 \\ 1 & 0 \end{pmatrix}\begin{pmatrix} x \\ y \end{pmatrix} = \begin{pmatrix} y \\ x \end{pmatrix}$$
$$A_2(C_{2a})z = -z$$

Thus, if we associate the polarization vectors P_x, P_y and P_z with the spinors $\bar{s}_x, \bar{s}_y$ and $\bar{s}_z$ using the formulas (1.8.24), (1.8.25) and (1.8.26), then the spinors transformed by the double-valued IR $E'(C_{2a})$ will correspond to the polarization vectors P'_x, P'_y and P'_z, obtained by transforming the original polarization vector using the matrix of single-valued IR of point group. (In this particular case of D_4 group, the components of the vector are transformed by two IRs $E(C_{2a})$ and $A_2(C_{2a})$.)

1.9 Direct Products of Groups and Representations

Let us define direct product of groups $G_1 \times G_2$ as a set of pairs of elements (r_1, s_2), where $r_1 \in G_1$ $s_2 \in G_2$, with the group law of multiplication:

$$(r_1, s_2) \times (p_1, q_2) = (r_1 p_1, s_2 q_2) \tag{1.9.1}$$

It is easy to see that the group postulates hold for the direct product, if the identity element is defined as (E_1, E_2) and the inverse element as $(r_1, s_2)^{-1} = (r_1^{-1}, s_2^{-1})$. If a system consists of two independent parts with symmetries G_1 and G_2, then the total symmetry group of this system is a direct product $G_1 \times G_2$. For example, if the Hamiltonian has symmetry G and there is no spin-orbit interaction, then the complete symmetry group of the total wavefunction, consisting of orbital and spin parts, is a direct product $G \times G$.

In some cases, the symmetry group G can be represented as a direct product of its subgroups $G_1 \times G_2$. This is only possible if the elements of the subgroups commute with each other. For example, inversion commutes with all rotations, so all groups containing inversion are represented as a direct product of a group of pure rotations and a group C_i consisting of two elements E and I. All point groups containing inversion can be represented as direct products of subgroup of pure rotations by the group $\{E, I\}$ and all IRs are divided into two classes: even and odd, what are denoted by subscripts g and u, respectively. The rotation group $\mathsf{O}^+(3)$ can be expanded by multiplying by group $\{E, I\}$. In this case, IRs D^l of $\mathsf{O}^+(3)$ can be expanded by multiplication by 1 or -1 and are written as D^{l+} or D^{l-}. However, since the spherical functions with even $l-$ values are even and spherical functions with odd $l-$values are odd, the superscripts $+$ and $-$ are usually dropped.

Consider two electrons with angular momentums l_1 and l_2 in a centrally symmetrical potential. If the interaction between them can be neglected, then the symmetry group of this system is the direct product of two rotation groups $\mathsf{O}(3) \times \mathsf{O}(3)$. Each one of the electron's wavefunction will be transformed according to the IR D^l of the rotation group. Total wavefunction of two electrons is a direct product of two wavefunctions and transforms according to a direct product of representations $D^{l_1} \times D^{l_2}$. If rotations in two spaces are independent, this representation is irreducible. The character of IR D^l for the rotation by the angle φ is given by the formula:

$$\chi^l(\varphi) = \sum_{m=-l}^{l} e^{im\varphi} \tag{1.9.2}$$

Trace of the direct product of matrices equals to the product of traces:

$$\chi^{l_1 \times l_2}(\varphi_1, \varphi_2) = \sum_{m=-l_1}^{l_1} e^{im\varphi_1} \sum_{m=-l_2}^{l_2} e^{im\varphi_2} \tag{1.9.3}$$

In a coupled system, the rotation angles are not independent and must be the same. In this case, the character of the product is decomposed into the characters of IRs of the rotation group:

$$\chi^{l_1 \times l_2}(\varphi) = \sum_{m_1=-l_1}^{l_1} e^{im_1\varphi} \sum_{m_2=-l_2}^{l_2} e^{im_2\varphi} = \sum_{m_1=-l_1}^{l_1} \sum_{m_2=-l_2}^{l_2} e^{i(m_1+m_2)\varphi} = \tag{1.9.4}$$

$$\sum_{L=|l_1-l_2|}^{l_1+l_2} \sum_{M=-L}^{L} e^{iM\varphi} = \sum_{L=|l_1-l_2|}^{l_1+l_2} \chi^L(\varphi) \tag{1.9.5}$$

Thus making use of the group character technique we obtained the rule of angular momentum coupling, which is known from quantum mechanics.

Now consider the general case of a system of two non-interacting particles with symmetry group G. Let the first particle have a wavefunction $\psi_i^\mu (i = 1, ...n_\mu)$, belonging to IR D^μ, and the second is the wavefunction $\varphi_j^\nu (j = 1, ...n_\nu)$, belonging to IR D^ν. The wavefunction of such a system is equal to the direct product of the wavefunctions of the parts, i.e. it is written as

$$\Psi_{ij}^{\mu\times\nu} = \psi_i^\mu \varphi_j^\nu \; i = 1, ...n_\mu, \; j = 1, ...n_\nu \tag{1.9.6}$$

Under the action of the elements (g_1, g_2) of the direct product $G \times G$ of the symmetry groups of individual particles, the wavefunction of the system is transformed by a direct product of the matrices of the initial IRs of G:

$$(g_1, g_2)\Psi_{ij}^{\mu\times\nu} = \sum_{kl} D_{ik}^\mu(g_1) D_{jl}^\nu(g_2) \psi_k^\mu \varphi_l^\nu \tag{1.9.7}$$

Thus, the matrix of the representation of direct product is given by the formula:

$$D_{ij,kl}^{\mu\times\nu}(g_1, g_2) = D_{ik}^\mu(g_1) D_{jl}^\nu(g_2) \tag{1.9.8}$$

Summing up the diagonal elements, we get that the character of the direct product of two representations is equal to the product of their characters:

$$\chi^{\mu\times\nu}(g_1, g_2) = \chi^\mu(g_1)\chi^\nu(g_2) \tag{1.9.9}$$

When taking four IRs D^μ, D^ν, D^λ and D^κ of group G it is easy to obtain the orthogonality relations for the direct product of groups $G \times G$:

$$\frac{1}{|G|^2} \sum_{g_1, g_2 \in G} \chi^\mu(g_1)\chi^\nu(g_2)\chi^{\lambda *}(g_1)\chi^{\kappa *}(g_2) = \frac{1}{|G|} \sum_{g_1 \in G} \chi^\mu(g_1)\chi^{\lambda *}(g_1) \frac{1}{|G|} \times, \left[\sum_{g_2 \in G} \chi^\nu(g_2)\chi^{\kappa *}(g_2) \right] = \delta(\mu, \lambda)\, \delta(\nu, \kappa) \tag{1.9.10}$$

It follows from this relation that the character of a direct product $G \times G$ given by the formula (1.9.9) is irreducible in the direct product of groups. However another case is also possible. If the two parts of the system under consideration are connected, then the elements of the group G must act on them in the same way, i.e. the symmetry group of the system is the set (g_i, g_j), where $i = j$. In this case, the direct product of representations decomposes into the sum of the IRs of the group G:

$$D^\mu \times D^\nu = \sum_\lambda f_\lambda D^\lambda \tag{1.9.11}$$

The basis functions $\zeta_{m\gamma}^\lambda$ of the representations on the right-hand side are expressed as linear combinations of the original wavefunctions:

$$\zeta^{\lambda}_{m\gamma} = \sum_{i,j} \langle \mu i, \nu j | \lambda m \gamma \rangle \, \psi^{\mu}_{i} \varphi^{\nu}_{j} \tag{1.9.12}$$

where γ stands for additional quantum number, which is required if any IR D^{λ} appears twice or more in the decomposition of direct product. In the case of spherical symmetry, each IR occurs only once in the decomposition of direct product $D^{l_1} \times D^{l_2}$ and no additional quantum number is required. In the present chapter, additional quantum numbers will be dropped in formulas to be short. Additional quantum numbers appearing in the induction will be considered in Chaps. 3 and 4.

Unitary transformation coefficients (expressions in brackets in formula (1.9.12)) are called Clebsch-Gordan coefficients. The Clebsch-Gordan coefficients satisfy the following sum rules:

$$\sum_{i,j} \langle \mu i, \nu j | \lambda m \rangle \langle \mu i, \nu j | \lambda' m' \rangle = \delta \left(\lambda, \lambda' \right) \delta \left(m, m' \right) \tag{1.9.13}$$

$$\sum_{\lambda,m} \langle \mu i, \nu j | \lambda m \rangle \langle \mu i', \nu j' | \lambda m \rangle = \delta \left(i, i' \right) \delta \left(j, j' \right) \tag{1.9.14}$$

In the case of spherical symmetry, Clebsch-Gordan coefficients have a number of symmetry properties. In order to use these symmetry properties the $3j-$ symbols, related to the Clebsch-Gordan coefficients, are introduced by the following relation:

$$\langle j_1 j_2 m_1 m_2 | j m \rangle = (-1)^{-j_1+j_2-m} (2j+1)^{1/2} \begin{pmatrix} j_1 & j_2 & j \\ m_1 & m_2 & -m \end{pmatrix} \tag{1.9.15}$$

The magnetic quantum numbers of $3j-$symbol satisfy the following relation:

$$m_1 + m_2 + m = 0 \tag{1.9.16}$$

The $3j-$symbols have some symmetry properties, which are useful for consideration of terms of two equivalent electrons in atom. The $3j-$symbol does not change with a cyclic (even) permutation of its columns:

$$\begin{pmatrix} j_1 & j_2 & j \\ m_1 & m_2 & m \end{pmatrix} = \begin{pmatrix} j_2 & j & j_1 \\ m_2 & m & m_1 \end{pmatrix} = \begin{pmatrix} j & j_1 & j_2 \\ m & m_1 & m_2 \end{pmatrix} \tag{1.9.17}$$

When the permutation of columns is odd, as well as when the signs of the magnetic quantum numbers in the bottom row are changed, the $3j-$symbol is multiplied by the phase factor $(-1)^{j_1+j_2+j}$.

$$\begin{pmatrix} j_1 & j_2 & j \\ m_1 & m_2 & m \end{pmatrix} = (-1)^{j_1+j_2+j} \begin{pmatrix} j_2 & j_1 & j \\ m_2 & m_1 & m \end{pmatrix} = (-1)^{j_1+j_2+j} \begin{pmatrix} j_1 & j_2 & j \\ -m_1 & -m_2 & -m \end{pmatrix}$$

Two orthogonality relations for $3j-$symbols are written as

$$\sum_{j,m} (2j+1) \begin{pmatrix} j_1 & j_2 & j \\ m_1 & m_2 & m \end{pmatrix} \begin{pmatrix} j_1 & j_2 & j \\ m'_1 & m'_2 & m \end{pmatrix} = \delta_{m_1,m'_1} \delta_{m_2,m'_2} \tag{1.9.18}$$

$$\sum_{m_1 m_2} \begin{pmatrix} j_1 & j_2 & j \\ m_1 & m_2 & m \end{pmatrix} \begin{pmatrix} j_1 & j_2 & j' \\ m_1 & m_2 & m' \end{pmatrix} = \frac{\delta_{j,j'}\delta_{m,m'}}{(2j+1)} \tag{1.9.19}$$

1.10 Symmetrized Square of Representation. Wavefunction of Two Equivalent Electrons

In the case of two equivalent electrons, the direct product of the wavefunctions contains the same pairs of factors $\varphi_i\varphi_j$ and $\varphi_j\varphi_i$ $(i \neq j)$, which are transformed in the same way under the action of the element R of the symmetry group. We ascribe to the first basis set the electronic coordinate ξ_1 and to the second basis set the electronic coordinate ξ_2 and write these numbers as superscripts. Then the contribution of these two products to the kl row of the direct product, which we denote as Φ_{kl}, is written as

$$R\Psi_{kl} = \ldots D^{\nu}_{ki}(R)\ \varphi^1_i D^{\nu}_{lj}(R)\varphi^2_j + \ldots + D^{\nu}_{kj}(R)\ \varphi^1_j D^{\nu}_{li}(R)\varphi^2_i \tag{1.10.1}$$

It follows from this formula that the products $\varphi^1_i\varphi^2_j$ and $\varphi^1_j\varphi^2_i$ are transformed in the same way regardless of the order of the wavefunctions. Consider the case of a Kronecker square of IR D of dimension two with basis $\{\varphi_i\}$. The result of transformation of a direct product basis function $\left\{\Phi_{ij}\right\} = \{\varphi_i\} \times \{\varphi_i\}$ by the action of the operation R, i.e. $\Phi' = R\Phi$, may be written as

$$\Phi'_{11} = D_{11}D_{11}\varphi^1_1\varphi^2_1 + D_{11}D_{12}\varphi^1_1\varphi^2_2 + D_{12}D_{11}\varphi^1_2\varphi^2_1 + D_{12}D_{12}\varphi^1_2\varphi^1_2 \tag{1.10.2}$$

$$\Phi'_{12} = D_{11}D_{21}\varphi^1_1\varphi^2_1 + D_{11}D_{22}\varphi^1_1\varphi^2_2 + D_{12}D_{21}\varphi^1_2\varphi^2_1 + D_{12}D_{22}\varphi^1_2\varphi^2_2 \tag{1.10.3}$$

$$\Phi'_{21} = D_{21}D_{11}\varphi^1_1\varphi^2_1 + D_{21}D_{12}\varphi^1_1\varphi^2_2 + D_{22}D_{11}\varphi^1_2\varphi^2_1 + D_{22}D_{12}\varphi^1_2\varphi^2_2 \tag{1.10.4}$$

$$\Phi'_{22} = D_{21}D_{21}\varphi^1_1\varphi^2_1 + D_{21}D_{22}\varphi^1_1\varphi^2_2 + D_{22}D_{21}\varphi^1_2\varphi^2_1 + D_{22}D_{22}\varphi^1_2\varphi^2_2 \tag{1.10.5}$$

where the argument R in $D(R)$ is dropped for brevity.

Adding and subtracting the two middle rows, we get

$$\Phi'^{s}_{1} = \Phi'_{11} = (D_{11})^2\,\varphi^1_1\varphi^2_1 + D_{11}D_{12}\left(\varphi^1_1\varphi^2_2 + \varphi^1_2\varphi^2_1\right) + (D_{12})^2\,\varphi^1_2\varphi^2_2 \tag{1.10.6}$$

$$\Phi'^{s}_{2} = \Phi'_{12} + \Phi'_{21} = 2D_{11}D_{21}\varphi^1_1\varphi^2_1 + (D_{11}D_{22} + D_{12}D_{21})\left(\varphi^1_1\varphi^2_2 + \varphi^1_2\varphi^2_1\right) + 2D_{12}D_{22}\varphi^1_2\varphi^2_2 \tag{1.10.7}$$

$$\Phi'^{s}_{3} = \Phi'_{22} = (D_{12})^2\,\varphi^1_1\varphi^2_1 + D_{22}D_{21}\left(\varphi^1_1\varphi^2_2 + \varphi^1_2\varphi^2_1\right) + (D_{22})^2\,\varphi^1_2\varphi^2_2 \tag{1.10.8}$$

$$\Phi_1'^a = \Phi_{12}' - \Phi_{21}' = (D_{11}D_{22} - D_{12}D_{21})\left(\varphi_1^1\varphi_2^2 - \varphi_2^1\varphi_1^2\right) \qquad (1.10.9)$$

It follows from the above relations that the symmetrized basis $\left\{\varphi_1^1\varphi_1^2,\ \varphi_1^1\varphi_2^2 + \varphi_2^1\varphi_1^2,\ \varphi_2^1\varphi_2^2\right\}$ (first three formulas) and antisymmetrized basis $\left\{\varphi_1^1\varphi_2^2 - \varphi_2^1\varphi_1^2\right\}$ are transformed independently, i.e. the Kronecker square of IR of dimension two decomposes into symmetric and antisymmetric parts with respect to permutations of coordinates. The same conclusion can be proved for IR of any dimension of finite or continuous group. Denoting symmetrized and antisymmetrized square by square and curly brackets, respectively, we can write the result as follows:

$$D \times D = [D \times D] + \{D \times D\} \qquad (1.10.10)$$

Since the terms $i \neq j$ contribute to both sides and the term $i = j$ contributes only to the symmetrized square, the dimensions of the symmetrized and antisymmetrized squares are, respectively, $\frac{1}{2}n(n+1)$ and $\frac{1}{2}n(n-1)$. It is seen from formulas (1.10.6), (1.10.7) and (1.10.8) that the character of transformation matrix for symmetrized part is $(D_{11})^2 + (D_{11}D_{22} + D_{12}D_{21}) + (D_{22})^2$. Formula (1.10.9) represents the character of antisymmetrized square. Thus, the characters of the symmetrized and antisymmetrized squares may be written as

$$\chi\,[D \times D]\,(g) = \frac{\chi^2(g) + \chi(g^2)}{2} \qquad (1.10.11)$$

$$\chi\,\{D \times D\}\,(g) = \frac{\chi^2(g) - \chi(g^2)}{2} \qquad (1.10.12)$$

The same formulas are valid for IRs of finite groups of all dimensions and for continuous group. For projective IRs, these formulas are generalized as

$$\chi\left[\hat{D} \times \hat{D}\right](g) = \frac{\hat{\chi}^2(g) + \omega(g,g)\hat{\chi}(g^2)}{2} \qquad (1.10.13)$$

$$\chi\left\{\hat{D} \times \hat{D}\right\}(g) = \frac{\hat{\chi}^2(g) - \omega(g,g)\hat{\chi}(g^2)}{2} \qquad (1.10.14)$$

Example 1.4 Symmetrized and antisymmetrized squares of IRs T_{1u}, T_{2g} and E_g in O_h symmetry.

p-electrons in O_h symmetry belong to IR T_{1u} and the results of this example are relevant to atoms with p^2 configuration in crystal field. Also $d-$electrons in O_h symmetry belong to $T_{2g} + E_g$ and products of these IRs correspond to d^2-configuration in O_h symmetry. In the groups with inversion symmetry, the parity of a product of two odd IRs is obviously even and it is sufficient to consider subgroup of pure rotations O of O_h group. Characters of group O and symmetrized and antisym-

Table 1.18 IRs of group O and symmetrized and antisymmetrized squares of IRs E, T_1 and T_2. (In O_h group IRs of products are even for both even and odd IRs in the first column)

IR	E	$4C_3$	$4C_3^2$	$3C_2$	$3C_4$	$3C_4^3$	$6C_2'$	product
A_1	1	1	1	1	1	1	1	
A_2	1	1	1	1	−1	−1	−1	
E	2	−1	−1	2	0	0	0	
T_1	3	0	0	−1	1	1	−1	
T_2	3	0	0	−1	−1	−1	1	
$[T_1 \times T_1]$	6	0	0	2	0	0	2	$A_1 + E + T_2$
$\{T_1 \times T_1\}$	3	0	0	−1	1	1	−1	T_1
$[T_2 \times T_2]$	6	0	0	2	0	0	2	$A_1 + E + T_2$
$\{T_2 \times T_2\}$	6	0	0	−1	1	1	−1	T_1
$[E \times E]$	3	0	0	3	1	1	1	$A_1 + E$
$\{E \times E\}$	1	1	1	1	−1	−1	−1	A_2
$T_2 \times E$	6	0	0	−2	0	0	0	$T_1 + T_2$

metrized squares of IRs T_1, T_2 and E of product $T_2 \times E$ are presented in Table 1.18.

According to Pauli exclusion principle, total wavefunction is antisymmetric with respect to permutation of electronic coordinates. Hence, it follows that symmetrized squares of spatial parts correspond to singlet states and antisymmetrized squares of spatial parts correspond to triplet states. The product of states $E \times T_2$ may be singlet and triplet.

Example 1.5 Symmetry of singlet and triplet spin functions in O_h symmetry.
Characters of double-valued IRs of O_h groups and decompositions of their Kronecker squares are presented in Table 1.19. In O_h symmetry spin 1/2 function belongs to two-dimensional IR E_1' and angular momentum $j = 3/2$ belongs to four-dimensional IR G' (see Table 1.14). Formulas for symmetrized squares of projective IRs (1.10.13) and (1.10.14) include phase factors $\omega(g, g)$ for the squares of the group elements. These factors are also presented in Table 1.19. It follows from Table 1.19 that in O_h symmetry singlet spin function belongs to IR A_{1g} and triplet spin function belongs to IR T_{1g}. Other products presented in Table 1.19 will be discussed in subsequent examples.

According to the Pauli exclusion principle, the total wavefunction is antisymmetric with respect to permutations of the electron coordinates. In $L - S$ coupling scheme, an electron has two coordinated spin and orbital. Thus, if the total wavefunction is symmetrical with respect to permutation of spin (orbital) coordinates, it must be antisymmetrical with respect to permutation of orbital (spin) coordinates. Therefore, symmetrized square of the spatial part of the wavefunction must correspond to the antisymmetrized square of the spin part (singlet state). Also antisymmetrized square of the spatial part of the wavefunction must correspond to the symmetrized spin

Table 1.19 Characters of double-valued IRs of O_h group, projective factors $\omega(g, g)$ for the squared elements and decompositions of Kronecker products

O_h	E	$4C_3$	$4C_3^2$	$3C_2$	$3C_4$	$3C_4^3$	$6C_2'$	
$\omega(g, g)$	1	1	–1	–1	1	–1	–1	
E_1'	2	1	-1	0	$\sqrt{2}$	$-\sqrt{2}$	0	
E_2'	2	1	-1	0	$-\sqrt{2}$	$\sqrt{2}$	0	
G'	4	-1	1	0	0	0	0	
$\{E_1' \times E_1'\}$	1	1	1	1	1	1	1	A_{1g}
$[E_1' \times E_1']$	3	0	0	-1	1	1	-1	T_{1g}
$\{G' \times G'\}$	6	0	0	2	0	0	2	$A_{1g} + T_{2g} + E_g$
$E_1' \times G'$	8	-1	-1	0	0	0	0	$E_g + T_{1g} + T_{2g}$

part (triplet state). Consider the symmetry under permutations of the wave functions of two equivalent electrons in spherical symmetry. In this case, as is known from quantum mechanics, the state of each value of the angular momentum of two-electron function appears only once, so it is obvious that there is a rule for separating the spatial wavefunction Kronecker square into symmetric and antisymmetric parts.

Consider a sate of two equivalent electrons $j_1 = j_2 = j$, coupled in total momentum J. The direct product of the wavefunctions contains the terms $\psi_{jm_1}\psi_{jm_2}$ and $\psi_{jm_2}\psi_{jm_1}$, differing just by the numbers of electron coordinates. These terms contribute to one state Ψ_{JM} with coefficients equal to the Clebsch-Gordan coefficients

$$\langle jjm_1m_2|JM\rangle = (-1)^{-M}[J]^{1/2}\begin{pmatrix} j & j & J \\ m_1 & m_2 & -M \end{pmatrix}$$

and

$$\langle jjm_2m_1|JM\rangle = (-1)^{-M}[J]^{1/2}\begin{pmatrix} j & j & J \\ m_2 & m_1 & -M \end{pmatrix}.$$

These $3-j$ symbols differ by a permutation of columns followed by a multiplication by the factor $(-1)^{2j+J}$. For a given j this value may be negative or positive, depending on J. This means that the result of coupling of the angular momentums of two equivalent electrons consists of terms that are already symmetrized or antisymmetrized with respect to permutations of electronic coordinates.

For integer values of j, the quantity $2j$ is always even, and the parity of two-electron state equals to the parity of the value J. For half-integer j, the value $2j$ is always odd, so the parity of total state is opposite to the parity of the value J. If the total spin is zero (singlet state), then the spin wavefunction is odd, and if the spin is equal to one, then the spin wavefunction is even with respect to permutation.

From this we get the following rules for construction of the wavefunction of two equivalent electrons obeying the Pauli exclusion principle:

1. The orbital part of the singlet wavefunction must be symmetrical with respect to permutations, i.e. L must be an even number.

2. The orbital part of the triplet wavefunction must be antisymmetric with respect to permutations, i.e. L must be an odd number.

These two rules can be reduced to one: for two equivalent electrons in the $L-S$ coupling scheme: the value $L+S$ of the term ^{2S+1}L must be even.

The wavefunction in the $j-j$ coupling scheme must be antisymmetric, therefore, when coupling total angular momentums j of two equivalent electrons, the resulting angular momentum J must be even for a half-integer j of coupled electrons.

If the electrons with the same $l-$ values are non-equivalent, their principal quantum numbers are different. In this case for all values of the total angular momentum L triplet and singlet states are possible.

Example 1.6 Find the wavefunction of two equivalent $p-$ electrons in a field of O_h symmetry, taking into account the spin-orbit interaction.

The task can be solved in various ways, depending on the relative strength of interactions, namely, Coulomb intra-atomic interaction U_{Cul}, spin-orbit coupling U_{SO} and crystal field interaction U_{CF}.

From this we get the following solutions of the problem.

Solution 1 $U_{Cul} > U_{SO} > U_{CF}$. When coupling two angular momentums $l=1$, we obtain total angular momentums $L=2,1,0$. Since the total wavefunction is antisymmetric with respect to permutations of electron coordinates, the sum $L+S$ must be even, i.e. only terms ^1D, ^3P and ^1S are possible. Adding the orbital angular momentum with triplet spin, we obtain the following values of total angular momentum $J=2,2,1,0,0$. The value of the angular momentum is uniquely related to parity only in the case of one electron, since spherical functions with an even value of l are even with respect to the inversion, and functions with an odd value of l are odd. The parity of a many-electron wavefunction is $(-1)^{lN}$. In our case, $l=1$ and $N=2$, and the spatial part of the wavefunction is even. The spin function is always even. Using the reduction scheme presented in Table 1.9, we obtain a complete set of IRs in O_h symmetry: $2T_{2g}+2E_g+T_{1g}+2A_{1g}$.

Solution 2 In a strong spin-orbit coupling case three atomic two-electron configurations are possible: $p_{1/2}^2$, $p_{3/2}^2$ and $p_{1/2}p_{3/2}$. In the first and in the second cases, electrons are equivalent, and to satisfy Pauli exclusion principle resultant angular momentums should be even. Thus, we obtain $J=0$ for the first case and $J=0,2$ for the second case. In the third case, electrons are non-equivalent and all states are possible, $J=2,1$. Thus, the set of total of angular momentums is the same as in previous solution and we obtain the same set of IRs of O_h group.

Solution 3 Now consider the case of a large crystal field and small Coulomb intra-atomic interaction $U_{CF} > U_{Cul} > U_{SO}$. Atomic $p-$ functions belong to IR T_{1u} of O_h group. Making use decompositions of Kronecker squares of T_{1u}, presented in Table 1.18, we obtain following states $^1T_{2g}+^1E_g+^1A_{1g}+^3T_{1g}$, where left superscript denotes spin multiplicity. It is seen in Table 1.19 that singlet and triplet spin functions belong to IRs A_{1g} and T_{1g}, respectively. Hence, spin-orbit coupling doesn't change the symmetry of three singlet states. For the triplet state, we must decompose the direct product $T_{1g} \times T_{1g}$. This product equals to the sum of symmetrized and antisymmetrized parts presented in Table 1.18. Thus, we obtain finally also $2T_{2g} + 2E_g + T_{1g} + 2A_{1g}$.

Solution 4 $U_{SO} > U_{Cul} > U_{CF}$. In O_h field atomic functions $p_{1/2}$ and $p_{3/2}$ belong to double-valued IRs E_1' and G', respectively. Thus possible states belong to products $\{E_1' \times E_1'\} + \{G' \times G'\} + E_1' \times G'$. Making of data of Table 1.19 we obtain the same result as in previous solutions.

Example 1.7 Find possible triplet terms of configuration d^2 in O_h point group symmetry.

Making use of angular momentum coupling rules we obtain that possible angular moments are $L = 0, 1, 2, 3$ and 4. As it was pointed out above, it follows from Pauli exclusion principle that the $L + S$ values for the states of two equivalent electrons should be even. Thus, we obtain all possible terms of d^2 configuration: ${}^1S$, ${}^3P$, ${}^1D$, ${}^3F$ and ${}^1G$. According to Hund's rules the lowest energy has the term with the largest S value, and if there are some terms with the same S, the lowest energy has the term with the largest L. Thus in our case the ground term is ${}^3F$. However, since in crystal field angular momentum L is not a good quantum number we must consider also the term ${}^3P$. To obtain the crystal field splitting of ground terms, one can use Table 1.8 for the splitting of spherical functions. Note that the spatial parity of the spherical functions coincides with the parity of the angular momentum value and in O_h symmetry $p-$ and $f-$ functions belong to odd IRs. The parity of the l^N configuration equal to $(-1)^{lN}$ and all terms under consideration are even. Thus when using the data of Table 1.8 and replacing subscript "u" by "g", we obtain that possible symmetries of triplet states of d^2 configuration in cubic crystal are

$$2^3T_{1g} +^3 A_{2g} +^3 T_{2g}$$

In a strong crystal field $d-$ states are split into E_g and T_{2g} states. Triplet spin state is symmetric with respect to permutations and corresponding spatial parts belong to antisymmetrized squares of IRs. Since $d-$electrons of E_g and T_{2g} symmetry are not equivalent in crystal field, their product appears both in singlet and triplet states. Making use of data of Table 1.18 we can write

$$\{E_g \times E_g\} = A_{2g}$$

$$\{T_{2g} \times T_{2g}\} = T_{1g}$$

$$T_{2g} \times E_g = T_{1g} + T_{2g}$$

Thus for two-electron states in point group symmetry both methods give the same results.

Example 1.8 Find possible ground terms of configuration d^6 in T_d point group symmetry.

If crystal field is smaller than atomic Coulomb interaction, the atomic ground term ^5D of d^6 configuration splits into two terms ^5E and $^5T_2$ (see Table 1.8). Now suppose that crystal field is stronger, than atomic Coulomb interaction. As it is shown in Chap. 3, in T_d symmetry T_2 level is below than E. Thus six $d-$electrons occupy three spin-up and three spin-down T_2 levels and total spin of the state $(T_2)^6$ is zero.

1.11 Projection Operators and Wavefunctions in Molecules

Acting by the elements g of the group G on a wavefunction φ_1 we obtain the orbit of this function, which in the general case consists of $|G|$ elements. Under the action of elements g_i of the group, these $|G|$ functions $\{\varphi_i\}$ are transformed according to the regular representation of the group, which performs permutations of elements of the basis. This basis set is reducible in a general case and we need the method to find linear combinations belonging to the rows of IRs of the entire group. To find linear combinations, belonging to $k-$ row of IR of the whole group projection operators are used:

$$P^{\mu}_{kk} = \frac{n_{\mu}}{|G|} \sum_{g \subset G} D^{\mu *}_{kk}(g) g \tag{1.11.1}$$

where n_{μ} is the dimension of IR D^{μ}. The derivation of this formula is in the book of Hamermesh [2]. Then, taking the function φ_1 as the initial one, we obtain a linear combination that transforms along the k row of IR D^{μ}:

$$\psi^{\mu}_{kk} = P^{\mu}_{kk}\varphi_1 = \frac{n_{\mu}}{|G|} \sum_{g \subset G} D^{\mu *}_{kk}(g) g \varphi_1 \tag{1.11.2}$$

It should be noted that, if a subset of functions $\{\bar{\varphi}_i\}$ of dimension $\bar{n}$ is invariant with respect to a subgroup H, then the number of elements in the orbit will be $\bar{n}\,|G|\,/\,|H|$. In the general projection operator method, the existence of subset invariant in a subgroup does not affect the final formula. In symmetrical molecules H is a local symmetry group of atom and $\{\bar{\varphi}_i\}$ is a basis of atomic orbitals. It will be shown in Chap. 3 that the induced representation method makes it possible to simplify the orbital analysis in symmetrical molecules.

Example 1.9 Construct a group orbitals for a square complex in which $s-$orbitals are located at the vertices of the square (see Fig. 1.11). The square has the symmetry group C_{4v}, the IRs of which are given in Table 1.3.

The results of action of projection operators (1.11.2) corresponding to one-dimensional IRs on function s_1 are presented in Table 1.20. It is seen from this

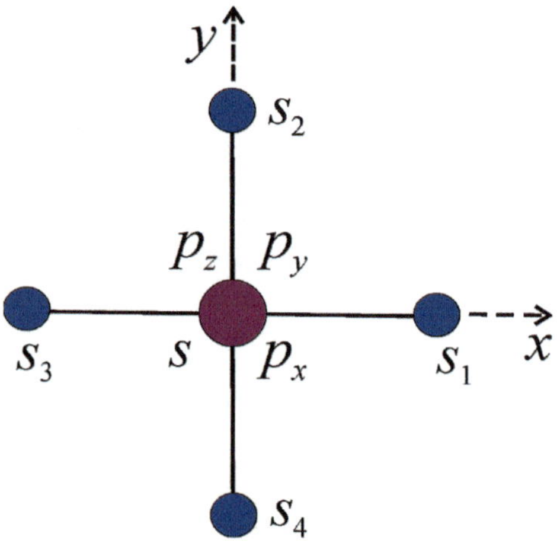

Fig. 1.11 Basis set for a square molecule, in which s, p_x, p_y and p_z are orbitals of the central atom, and s_1, s_2, s_3 and s_4 are $s-$orbitals of atoms forming a square

Table 1.20 Projecting of the basis set of $s-$ orbitals located at the vertices of the square (see Fig. 1.11) on one-dimensional IRs of the symmetry group C_{4v}

g	E	C_{2z}	C_{4z}	C_{4z}^3	σ_y	σ_x	σ_a	σ_b	Basis set
$P^{A_1}(g)s_1$	s_1	s_3	s_2	s_4	s_1	s_3	s_4	s_2	$\frac{1}{4}(s_1+s_2+s_3+s_4)$
$P^{A_2}(g)s_1$	s_1	s_3	s_2	s_4	$-s_1$	$-s_3$	$-s_4$	$-s_2$	0
$P^{B_1}(g)s_1$	s_1	s_3	$-s_2$	$-s_4$	s_1	s_3	$-s_4$	$-s_2$	$\frac{1}{4}(s_1-s_2+s_3-s_4)$
$P^{B_2}(g)s_1$	s_1	s_3	s_2	s_4	$-s_1$	$-s_3$	$-s_4$	$-s_2$	0

table that there are basis sets of A_1 and B_1 symmetry, but the projections on A_2 and B_2 vanish.

Moving on to two-dimensional IR E, note that since the formula for the projection operator (1.11.2) contains only diagonal elements, it suffices to sum only over those elements of the group whose matrices have nonzero elements on the diagonal, namely, E, C_{2z}, σ_x, σ_y (see Table 1.3). The results are shown in Table 1.21. It is seen from Table 1.21 that when projecting of s_1 on the first row we obtain $\frac{1}{2}(s_1 - s_3)$, but the projection on the second row vanishes. To obtain the function of the second row we need to start projection from the function s_2.

Suppose that there are $s-$ and $p-$orbitals on the central atom. Making use of IRs of C_{4v} presented in Table 1.3 it is immediately verified that $s-$ and p_z-orbitals belong to IR A_1 and p_x- and p_y- orbitals belong to the first and to the second rows of IR E, respectively. Since the Hamiltonian is scalar, only orbitals belonging to the same IR can interact. Thus, we see that $s-$ and p_z-orbitals of the central atom interact with A_1 combination of $s-$orbitals of ligands, and p_x- and p_y-orbitals of the central atom interact with E linear combinations of ligand's $s-$orbitals. Linear

Table 1.21 Projecting the basis set of $s-$orbitals located at the vertices of the square (see Fig. 1.11) on two-dimensional IR E of group C_{4v}

g	E	C_{2z}	σ_x	σ_y	Basis set
$P_1^E(g)s_1$	s_1	$-s_3$	$-s_3$	s_1	$\frac{1}{2}(s_1 - s_3)$
$P_2^E(g)s_1$	s_1	$-s_3$	$-s_3$	$-s_1$	0
$P_1^E(g)s_2$	s_2	$-s_4$	$-s_2$	s_4	0
$P_2^E(g)s_2$	s_2	$-s_4$	s_2	$-s_4$	$\frac{1}{2}(s_2 - s_4)$

combination of $s-$orbitals of ligands of B_1 symmetry doesn't interact with the central atom in this case.

Acknowledgements The author acknowledges Saint-Petersburg State University for a research project 116812735 ≪ New quantum materials and topological nanosystems based on spin-orbit interaction and magnetism for application in modern spintronics and quantum computing ≫

Chapter 2
Magnetic Groups and Their Applications

Time inversion symmetry has no far-reaching consequences in the theory of atomic spectra. It provides a much more powerful tool in the investigation of systems of lower symmetry...

E.Wigner [5]

Abstract The time-reversal symmetry is very useful in considerations of the systems of low symmetry and of magnetic structures. Magnetic (Shubnikov) groups and Wigner theory of their corepresentation are considered in the present chapter. More emphasis is placed on the examples of the construction of corepresentations of magnetic groups. This chapter is introductory for Chaps. 3 and 4.

2.1 Time-Reversal Operation and Magnetic Group Symmetry

Along with the geometric elements of symmetry, a quantum system has symmetry with respect to time reversal. In quantum mechanics, time reversal ϑ is understood as reversal of the direction of motion.

Consider the action of the time-reversal operation ϑ on the wave function whose evolution in time is determined by the Schrödinger equation. Let ψ_k be eigenfunction and E_k be eigenvalue of the time-independent Schrödinger equation $\hat{H}\psi = E\psi$. Then the evolution of the wavefunction of the system in time is written as

$$\psi_k(t) = \exp(-iE_k t/\hbar)\psi_k \tag{2.1.1}$$

The action of time reversal ϑ on this function results in complex conjugation:

$$\vartheta\psi_k(t) = \exp(iE_k t/\hbar)\vartheta\psi_k$$

V. G. Yarzhemsky, *Novel Group Theoretical Methods for Electron Structure Theory*,
https://doi.org/10.1007/978-3-031-72802-0_2

With free motion, the complex conjugation results in a change of the direction of motion, i.e. the sign of $k = p/\hbar$:

$$\vartheta e^{ikr} = e^{-ikr}$$

Let us determine the result of the action of the time-reversal operator ϑ on the complex coefficients c_k.

Since ϑ is a symmetry operator, it leaves the norm of the wavefunction unchanged:

$$\left(\psi_k^*, \psi_k\right) = \left(\vartheta\psi_k^*, \vartheta\psi_k\right) \tag{2.1.2}$$

The propagation in time (2.1.1) doesn't change this relation. Also it is essential to determine the result of the action of the time-reversal operator on the complex factors c_k before the wavefunction. It is immediately verified that (2.1.1) is fulfilled if $\vartheta c_k = c_k^*$, i.e. the time-reversal operator should be antilinear. An antilinear operator that satisfies this condition is called antiunitary. Any antiunitary operator can be represented as a product of the unitary operator U and the complex conjugation operator K:

The time-reversal operator applied to ψ_k twice results in a function that differs from the original one by a constant factor c only. Let us show that this factor can only be equal to 1 and -1. The operator ϑ^2 can be written as

$$\vartheta^2 = UKUK = UU^* = cE$$

where E is the identity matrix.

Due to the unitarity of the operator U, it can be substituted into this formula that $E = UU^\dagger$. Then we get that $U^* = cU^\dagger$, or $U = cU^T$. Passing to the transposed operators in the last equality, we get $U^T = cU$. Substituting U^T from the last equality to the previous one we obtain a relation $U = c^2U$, from which we deduce finally that $c = \pm 1$, or:

$$\vartheta^2 = \pm 1 \tag{2.1.3}$$

The upper sign corresponds to spinless wavefunctions, or an even number of fermions, and the lower sign to an odd number of fermions.

Different physical quantities behave differently when reversing time (reversing the direction of motion). The coordinates, potential, and kinetic energy either do not depend on the direction of motion or contain an even degree of time variable. Time reversal commutes with such operators, in particular, with the Hamiltonian:

$$\vartheta\hat{H} = \hat{H}\vartheta \tag{2.1.4}$$

For the momentum operator, we have

$$UK\left(-i\hbar\frac{\partial\varphi}{\partial x}\right) = i\hbar UK\frac{\partial\varphi}{\partial x}$$

Thus, the action of the time-reversal operator on the wavefunction of free motion leads to a change in the direction of motion and for momentum operator $\hat{p}$, we can write

$$\vartheta\hat{p} = -\hat{p}\vartheta \tag{2.1.5}$$

In quantum mechanics without taking into account the spin, the time-reversal operator is equivalent to complex conjugation:

$$\vartheta\varphi = \varphi^*$$

Let us consider the expression of the time-reversal operator in the basis of spin operators. Spin operators, as momentum operators, must anticommute with time reversal. The spin operator matrices s_x and s_z are real and commute with complex conjugation, so the unitary matrix U must anticommute with s_x and s_z. On the other hand, the complex spin matrix s_y anticommutes with complex conjugation and, therefore, must commute with the matrix U. As is known, the only matrix commuting with s_y and anticommuting with s_x and s_z is the matrix s_y. Thus, we obtain that in the theory that takes into account the spin, the time-reversal operator is equal to the product of the spin matrix $-is_y$ and the complex conjugation:

$$\vartheta = -is_yK = \begin{pmatrix} 0 & -1 \\ 1 & 0 \end{pmatrix} K \tag{2.1.6}$$

Thus, the action of time-reversal operator on the spinors $\bar{s}_y$, $\bar{s}_y$ and $\bar{s}_z$ with spin projections $+1$ on the axes x, y and z, respectively (see Sect. 1.8) is written as

$$\vartheta\bar{s}_x = \begin{pmatrix} 0 & -1 \\ 1 & 0 \end{pmatrix} K\frac{1}{\sqrt{2}}\begin{pmatrix} 1 \\ 1 \end{pmatrix} = \frac{1}{\sqrt{2}}\begin{pmatrix} -1 \\ 1 \end{pmatrix} \tag{2.1.7}$$

$$\vartheta\bar{s}_y = \begin{pmatrix} 0 & -1 \\ 1 & 0 \end{pmatrix} K\frac{1}{\sqrt{2}}\begin{pmatrix} 1 \\ i \end{pmatrix} = \frac{1}{\sqrt{2}}\begin{pmatrix} i \\ 1 \end{pmatrix} \tag{2.1.8}$$

$$\vartheta\bar{s}_z = \begin{pmatrix} 0 & -1 \\ 1 & 0 \end{pmatrix} K\begin{pmatrix} 1 \\ 0 \end{pmatrix} = \begin{pmatrix} 0 \\ 1 \end{pmatrix} \tag{2.1.9}$$

Making use of formulas (1.8.24), (1.8.25) and (1.8.26), it is immediately verified that the time-reversal changes the sign of projections of these spinors on the axes x, y and z, respectively, and we can write

$$\vartheta \uparrow = \downarrow, \tag{2.1.10}$$

where arrow denoted spin projection on any axis.

Also when acting by ϑ twice on a spinor $\bar{s}_y$ we obtain, in agreement with (2.1.3), that

$$\vartheta^2\bar{s}_y = \begin{pmatrix} 0 & -1 \\ 1 & 0 \end{pmatrix} K \frac{1}{\sqrt{2}} \begin{pmatrix} i \\ 1 \end{pmatrix} = \frac{1}{\sqrt{2}} \begin{pmatrix} -1 \\ -i \end{pmatrix} = -\bar{s}_y \tag{2.1.11}$$

The action of the time-reversal operator on the electron function in a crystal $e_{\uparrow}^{i(k \cdot r)}$, which is represented as a plane wave and spin-up function $\bar{s}_z$, is written as

$$-is_y K e_{\uparrow}^{i(k \cdot r)} = \begin{pmatrix} 0 & -1 \\ 1 & 0 \end{pmatrix} K e^{i(k \cdot r)} \begin{pmatrix} 1 \\ 0 \end{pmatrix} = e^{-i(k \cdot r)} \begin{pmatrix} 0 \\ 1 \end{pmatrix} = e_{\downarrow}^{-i(k \cdot r)} \tag{2.1.12}$$

Thus we proved a Kramers' theorem: the states $e_{\uparrow}^{i(k \cdot r)}$ and $e_{\downarrow}^{-i(k \cdot r)}$ are degenerated due to time-reversal symmetry. Note that if the crystal has a symmetry with respect to inversion I, the states k and $-k$ are degenerated due to spatial symmetry and the states $\uparrow$ and $\downarrow$ are degenerated due to time-reversal symmetry. In this case, there are four degenerated states at general k. Kramers' theorem means also that if a crystal hasn't inversion symmetry I, the states $e_{\uparrow}^{i(k \cdot r)}$ and $e_{\downarrow}^{-i(k \cdot r)}$ are degenerated due to time-reversal symmetry.

Point symmetry groups of molecules and space groups of crystals describe external symmetry, which does not depend on the properties of the internal electronic structure. One of the most important of these properties of the electronic structure is the direction of the magnetic moment. Usually in group theory, the operation of changing the magnetic moment is called time reversal. In ferromagnetic materials, the magnetic moments are oriented in one direction. Time reversal changes the direction of the magnetic moment and is not an element of spatial symmetry. In this case, the system has only a point group symmetry G. If all directions of the magnetic moment are equally probable at each point of the crystal, then time reversal does not change anything. In this case, the system is said to be symmetric with respect to time reversal ϑ and the symmetry group is $G + \vartheta G$. In addition, one can consider symmetry groups in which some of the geometric elements are included in combinations with time reversal (reversal of the magnetic moment). The direction of the magnetic moment (electron spin) is conveniently depicted using color. Let's match the spin pointing up with white color, and the spin pointing down with black color. If we color parts of geometric figures in a certain way in white and black, then by the operation of symmetry we mean an operation that not only transforms the figure into itself, but also preserves the color of its parts. If the system has magnetic symmetry, i.e. time reversal is the symmetry element, then we will assume that the corresponding geometric figure is painted grey. Magnetic point groups include 32 point groups:

$$M = G \tag{2.1.13}$$

which are called Shubnikov groups of type-I.

Shubnikov groups of type-II (grey groups) include elements of point symmetry, and their time-reversal products:

$$M = G + \vartheta G \quad (2.1.14)$$

Magnetic groups of type-III (black-and-white) group are written as

$$M = H + \vartheta(G - H) \quad (2.1.15)$$

where H is a halving subgroup of point group G.

It follows from this definition that the number of black-and-white Shubnikov groups that can be obtained from a given point group is equal to the number of halving subgroups contained in it, which are different in the abstract sense. A complete set of Shubnikov groups of type-III corresponding to a given point group can be obtained directly from the analysis of character tables. To do this, we need to consider real one-dimensional representations of the given point group. The identity representation corresponds to the point group itself, and the remaining real one-dimensional representations define Shubnikov groups of type-III. The elements of the group G for which the characters of the given IR are positive for the group H, and the remaining elements of the group G are included together with the color change operation. The total number of magnetic groups obtained in this way is equal to the number of one-dimensional IRs with characters 1 and -1 different in the abstract sense. Two IRs are called different in the abstract sense, if they cannot be transformed into each other by a permutation of the axes. Having carried such consideration for all 32 point groups, it was obtained that the total number of point black-and-white Shubnikov groups is 58.

Example 2.1 Find magnetic groups, originating from point group D_{2h}.

It is seen in Table 2.1 that all IRs of D_{2h} $A_{g(u)}$, $B_{1g(u)}$, $B_{2g(u)}$ and $B_{3g(u)}$ are one-dimensional. Halving subgroups corresponding to IRs of $B_{ig(u)}$−type are equivalent in abstract sense, since they differ by permutation of axes x, y and z. Thus there are three magnetic groups distinct in an abstract since originating from IRs A_u, B_{1u} and B_{1g}. These groups, denoted as $m'm'm'$, mmm' and $m'm'm$, are shown in Fig. 2.1a, b and c, respectively.

Thus, there are three magnetic groups, namely, $m'm'm'$, mmm' and $m'm'm$ with unitary subgroups D_2, C_{2v} and C_{2h}. These Shubnikov groups are shown in Fig. 2.1.

It is seen from Fig. 2.1a, b and c that antiunitary operation (change of the color) is connected with the elements $\{I, \sigma_x, \sigma_y, \sigma_z\}$, $\{I, \sigma_z, C_{2x}, C_{2y}\}$ and $\{\sigma_x, \sigma_y, C_{2x}, C_{2y}\}$ for groups $m'm'm'$, mmm' and $m'm'm$, respectively.

The symmetry of a point group is related to the orientation of the spins of atoms, and for each point group one can define a Shubnikov group for which the system can have a total nonzero magnetic moment. Consider a crystal, consisting of atoms of the same type with a nonzero total spin (magnetic moment), and choose for

Table 2.1 Irreducible representations of group D_{2h}

	E	C_{2z}	C_{2y}	C_{2x}	I	σ_z	σ_y	σ_x
A	1	1	1	1	1	1	1	1
A_u	1	1	1	1	−1	−1	−1	−1
B_{1g}	1	1	−1	−1	1	1	−1	−1
B_{1u}	1	1	−1	−1	−1	−1	1	1
B_{2g}	1	−1	1	−1	1	−1	1	−1
B_{2u}	1	−1	1	−1	−1	1	−1	1
B_{3g}	1	−1	−1	1	1	−1	−1	1
B_{3u}	1	−1	−1	1	−1	1	1	−1

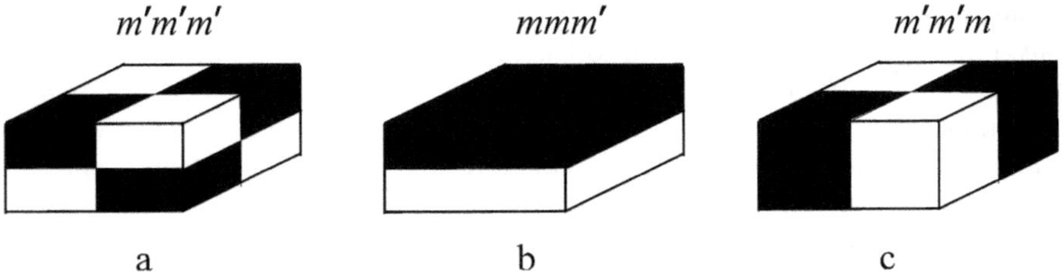

Fig. 2.1 Shubnikov groups arising from point group D_{2h}

definiteness the z-axis as the spin quantization axis. Rotations around the $z-$axis and inversion do not change the direction of the spin. Rotations around horizontal axes and reflections in vertical planes change the direction of the spin. If the point group contains elements of both types, then the magnetic moments of neighboring atoms are oriented opposite to each other and form an antiferromagnetic structure that does not have a magnetic moment. If the Shubnikov group is such that operations of the point group that change the direction of the spin are included together with the time-reversal operation ϑ, then all the spins are oriented in the same way and the crystal has a total magnetic moment. This is illustrated in Fig. 2.2 for groups D_{2h} and $m'm'm$. Figure 2.2 a shows spins in sectors of cuboid obtained by the actions of the elements of group D_{2h} on the spin s_z of the initial sector. Since half of the elements of D_{2h} group, namely, σ_x, σ_y, C_{2x}, C_{2y} change a spin, total number of spin-up components equals to the total number of spin-down components and total spin equals zero in this antiferromagnetic orientation. Figure 2.2b shows the same for the magnetic group $m'm'm$, in which geometrical operations changing spin appear with time reversal. In this case, the spins in all sectors are oriented by the same way and total spin is nonzero in this ferromagnetic structure.

Shubnikov space groups can be obtained from the characters of IRs space groups in the same way that point Shubnikov groups are obtained from the characters of point groups.

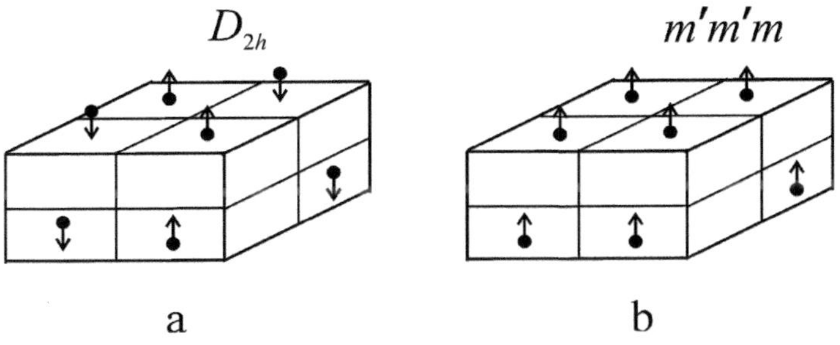

Fig. 2.2 Orientations of spins in groups D_{2h} and $m'm'm$

The first three are similar to the three types of Shubnikov point groups. The fourth type is due to the appearance of a new type of symmetry in space groups—color changes during translations. Thus, we have the following four types of Shubnikov space groups:

1. There is no antisymmetry operation. The Shubnikov group of type-I coincides with the Fedorov space group.

$$M = G \tag{2.1.16}$$

2. Shubnikov group of type-II or grey space groups. The crystal is symmetric with respect to the antisymmetry operation. The Shubnikov group is written as

$$M = G + \vartheta G \tag{2.1.17}$$

3. The Shubnikov space group of type-III (black-and-white group) is written as

$$M = H + \vartheta(G - H) \tag{2.1.18}$$

where H is a subgroup of index 2 of the Fedorov space group G, and $(G - H)$ does not contain pure translations. There are 674 such groups. Since the antisymmetry operator is not related to pure translations, these groups correspond to ordinary Bravais lattices.

The fourth type of Shubnikov space groups is associated with black-and-white Bravais lattices, which are obtained from ordinary Bravais lattices by splitting into black-and-white sublattices. From 14 regular Bravais lattices, 22 black-and-white lattices can be obtained.

4. The Shubnikov space group of type-IV is written as

$$M = G + \vartheta\{E|\tau\}G \tag{2.1.19}$$

Both the black-and-white sublattices have the same Fedorov symmetry group G, and $\{E|\tau\}$ is a translation that maps the black sublattice to white and vice versa. There are 517 such groups.

2.2 Corepresentations of Magnetic Groups

In the previous section, when describing Shubnikov groups, we used the term color change, which is associated with the antiunitary time-reversal operator (spin direction reversal). It should be noted that antiunitarity is not a property of Shubnikov groups, but appears when these groups are used to describe magnetic crystals.

When passing from unitary groups to antiunitary ones, which include, in addition to spatial operations, an antiunitary operation—complex conjugation or complex conjugation together with a spatial operation, the term irreducible representation (IR) is replaced by the term irreducible corepresentation (ICR). Let us consider the general properties of the ICRs, which follow from the properties of the transformation of wavefunctions by antiunitary operators.

Any of the considered magnetic groups M can be represented as a decomposition into left cosets with respect to the unitary subgroup G:

$$M = G + AG \tag{2.2.1}$$

Let a unitary subgroup G have a basis $|\psi_i\rangle$ of dimension n that is transformed by the IR D of the group G. Let the basis $|\psi_i\rangle$ be transformed into a basis $|\varphi_i\rangle$ of the same dimension under the action of an antiunitary element A. Consider the full basis transformation $|\psi_i, \varphi_i\rangle$ under the action of unitary R and S and antiunitary A and B elements.

The action of a unitary element R on $|\varphi_i\rangle$ c on can be represented as

$$R|\varphi_i\rangle = RA|\psi_i\rangle = A(A^{-1}RA)|\psi_i\rangle$$

Since $A^{-1}RA$ belongs to a unitary subgroup, we can use its representation by matrices D:

$$AD\left(A^{-1}RA\right)\psi_i = D^*\left(A^{-1}RA\right)A\psi_i = D^*\left(A^{-1}RA\right)\varphi_i$$

In writing the actions of antiunitary operator A on matrix D, we took into account that ϑ in antilinear operator.

Then the action of the unitary element on the complete basis can be written as

$$R\begin{pmatrix}\psi\\ \varphi\end{pmatrix} = \begin{pmatrix}D(R) & 0\\ 0 & D^*\left(A^{-1}RA\right)\end{pmatrix}\begin{pmatrix}\psi\\ \varphi\end{pmatrix} \tag{2.2.2}$$

In the following text, we will use the notation:

$$\bar{D}(R) = D^*(A^{-1}RA) \tag{2.2.3}$$

Any antiunitary element B can be represented as $B = AR$, where A is a fixed representative of the antiunitary left coset. Then the action of the antiunitary element on the unitary part of the basis can be written as

$$B|\psi_i\rangle| = AR|\psi_i\rangle = AD(R)|\psi_i\rangle = D^*(R)|\varphi_i\rangle = D^*(A^{-1}B)|\varphi_i\rangle$$

Similarly, the action of the antiunitary element on the antiunitary part of the basis can be represented as

$B|\varphi_i\rangle = BA|\psi_i\rangle = D(BA)|\psi_i\rangle$

Thus, the action of the antiunitary element on the full basis can be written as

$$B\begin{pmatrix}\psi\\ \varphi\end{pmatrix} = \begin{pmatrix}0 & D(BA)\\ D^*(A^{-1}B) & 0\end{pmatrix}\begin{pmatrix}\psi\\ \varphi\end{pmatrix} \tag{2.2.4}$$

The obtained formulas determine irreducible representation Δ of the magnetic group M, obtained from the IR D of its unitary subgroup G.

It should be noted that the usual multiplication rule for group representations holds for representations only if the first of the factors belongs to a unitary subgroup:

$$\Delta(R)\Delta(S) = \Delta(RS) \tag{2.2.5}$$

$$\Delta(R)\Delta(B) = \Delta(RB), \tag{2.2.6}$$

where $R, S \in G$ and $B \in AG$.

If the first factor is antiunitary, then the second factor enters with the sign of complex conjugation:

$$\Delta(B)\Delta^*(R) = \Delta(BR) \tag{2.2.7}$$

$$\Delta(B)\Delta^*(C) = \Delta(BC)) \tag{2.2.8}$$

where $R \in G$ and $B, C \in AG$.

Reducibility and irreducibility of antiunitary representations are defined similarly to those definitions for ordinary representations. If a basis set $|\psi_i, \varphi_i\rangle$ of a representation Δ can be transformed by a unitary transformation U into a direct sum of two basis sets that are invariant under all transformations from M, then the representation Δ is said to be reducible.

To find irreducible representations of Shubnikov groups, it is necessary to find out in which cases the representation obtained by the method considered above will be irreducible, and in which cases it will be reducible. As we shall see, this is determined by the relation between the representation D of the unitary group G and its conjugate representation:

$\overline{D}(R) = D^*(A^{-1}RA).$

Applying the conjugation operation to both sides of this relation, we obtain

$\overline{\overline{D}}(R) = \overline{D}^*(A^{-1}RA) = D(A^{-2}RA^2) = D^{-1}(A^2)D(R)D(A^2)$

This relation means that the doubly conjugated representation is equivalent to the original one. Therefore, a representation of the group M obtained from the IR

D is equivalent to a representation obtained from $\bar{D}$. The representations D and $\bar{D}$ themselves may or may not be equivalent. Therefore, when the representation $\Delta(D)$ of the magnetic group is restricted to a unitary subgroup, we obtain either two non-equivalent representations D and $\bar{D}$ or one representation D twice. If D and $\bar{D}$ are not equivalent, then the representation $\Delta(D)$ is irreducible. If D and D are equivalent, then they are connected by a unitary matrix N as $D(R) = N\bar{D}(R)N^{-1}$, and the reducibility of $\Delta(D)$ depends on the matrix N.

There are three types of corepresentations of Shubnikov groups, denoted as (a), (b) and (c). These types are determined by the equivalence of the matrices $D(R)$ and $\bar{D}(R)$ and the property of the transformation connecting these matrices N if $D(R)$ and $\bar{D}(R)$ are equivalent.

Case (a) $D(R)$ and $\bar{D}(R)$ are equivalent. $D(R) = N\bar{D}(R)N^{-1}, NN^* = +D\left(A^2\right)$

$$\Delta(R) = D(R),\ \Delta(B) = \pm D(BA^{-1})N \tag{2.2.9}$$

Case (b) $D(R)$ and $\bar{D}(R)$ are equivalent. $D(R) = N\bar{D}(R)N^{-1}, NN^* = -D\left(A^2\right)$

$$\Delta(R) = \begin{pmatrix} D(R) & 0 \\ 0 & D(R) \end{pmatrix} \Delta(B) = \begin{pmatrix} 0 & -D(BA^{-1})N \\ D(BA^{-1})N & 0 \end{pmatrix} \tag{2.2.10}$$

Case (c) $D(R)$ and $\bar{D}(R)$ are inequivalent.

$$\Delta(R) = \begin{pmatrix} D(R) & 0 \\ 0 & \bar{D}(R) \end{pmatrix} \Delta(B) = \begin{pmatrix} 0 & D(BA) \\ D(BA^{-1}) & 0 \end{pmatrix} \tag{2.2.11}$$

The type of corepresentation is defined by the Herring criterion, which includes the sum over the antiunitary left coset

$$\sum_{B \in AG} \chi(D(B^2)) = \begin{cases} +\lvert G\rvert \text{ in case } (a) \\ -\lvert G\rvert \text{ in case } (b) \\ 0 \quad \text{in case } (c) \end{cases} \tag{2.2.12}$$

Note that when calculating $\chi D(B^2)$ one should take into account that $\vartheta^2 = 1$, for spinless case and $\vartheta^2 = -1$ for half-integer spin. In the case of grey Shubnikov group, i.e. when $A = \vartheta$ the above relations can be written as

$$\sum_{R \in G} \chi(D(R^2)) = \begin{cases} \nu \lvert G\rvert \text{ in case } (a) \\ -\nu \lvert G\rvert \text{ in case } (b) \\ 0 \quad \text{in case } (c) \end{cases} \tag{2.2.13}$$

where

$$\nu = \begin{cases} 1 \text{ for integer spin} \\ -1 \text{ for half-integer spin} \end{cases}$$

In the case of half-integer spin and double-valued IRs, the projective factors $\omega(R, R)$ should be used in (2.2.12). Also projective factors appear in non-symmorphic space groups. In order to represent Herring criterion in terms of characters of IRs of point groups, we denote antiunitary elements as ϑB. In this case, the Herring criterion may be written as

$$\nu\left(\vartheta^2\right) \sum_{\vartheta B \in AG} \chi(B^2)\omega\left(B, B\right) = \begin{cases} |G| \text{ (a)} \\ -|G| \text{ (b)} \\ 0 \text{ (c)} \end{cases} \tag{2.2.14}$$

where $\chi(B^2)$ is a character for the element B^2 of a unitary subgroup. In the case of space group $\omega\left(B, B\right)$ is a total projective factor of representation, which is a product of projective factors appearing in double group and due to non-symmorphic structure of a space group.

2.3 Examples of Applications of Magnetic Groups

In the case of rotational symmetry, the following relation holds for the IRs:

$$\overline{D}^J = N\left(D^J\right)^*$$

For integer J, the matrix N transforms D^J into its conjugate matrix and satisfies the relation $NN^* = 1$. In this case, $\vartheta^2 = 1$, what corresponds to the case of no spin or an even number of fermions. In the case of an odd number of fermions, $NN^* = -1$ and $\vartheta^2 = -1$. This means that in the case of rotational symmetry, regardless of the value of J, time-reversal symmetry does not lead to any additional degeneracy.

In the absence of external magnetic fields, the wavefunction of an atom in a crystal is invariant under time reversal, i.e. a reversal of the direction of the magnetic moment. Therefore, the crystallographic point groups must be supplemented by the operation of time reversal, and the complete symmetry is described by the Shubnikov type-II group (gray group) $G + \theta G$. In this case, $\bar{D}(R) = D^*(\vartheta^{-1}R\vartheta) = D^*(R)$.

In the absence of spin, i.e. for single-valued IRs $D(A^2) = D(\vartheta^2) = +1$. Therefore, for real single-valued IRs, always $\bar{D}(R) = D(R)$, and corepresentations are constructed according to type (a), and it is always possible to choose the matrix N so that it is identity. For point groups, multidimensional IRs can always be chosen to be real, and for them there is also no additional degeneracy due to symmetry with respect to time reversal. For one-dimensional complex IRs $\bar{D}(R) = D^*(R)$ and the corepresentations are constructed according to type (c) and there is an additional degeneracy of energy levels due to time-reversal symmetry.

In the presence of spin (with an odd number of fermions) $D(A^2) = D(\vartheta^2) = -1$. In this case, to determine the type of a representation, using the technique

described in the previous section, one obtains the following rules for constructing corepresentations, depending on the type of the double-valued IRs of the point group:
1. Non-degenerate, complex—type (c).
2. Non-degenerate real—type (b).
3. Degenerate with some complex characters—type (c)
4. Degenerate with real characters—type (a).

Example 2.2 Making use of tables of double-valued IRs construct corepresentations for groups C_i, C_2 and C_s. Characters for these group are presented in Tables 2.2 and 2.3.

(a) Group C_i.

Since $\omega(I, I) = 1$, we have

$$\sum_{R\in G} \chi(R^2)\omega(R, R) = 1 + 1 = 2.$$

Since for half-integer spin $\nu = -1$, this is the case (b) (2.2.13) and corepresentation on the unitary subgroup equals to twice the IR and is written as $\begin{pmatrix} A_1' & 0 \\ 0 & A_1' \end{pmatrix}$. Because the space inversion is not essential for spin, another ICR $\begin{pmatrix} A_2' & 0 \\ 0 & A_2' \end{pmatrix}$ is physically equivalent to the first one.

(b) Groups C_2 and C_s.

For groups C_2 and C_s, we have $\omega(C_2, C_2) = -1$ and $\omega(\sigma, \sigma) = -1$, respectively. Thus for both groups the sum (2.2.13) is $\sum_{R\in G} \chi(R^2)\omega(R, R) = 1 - 1 = 0$.

Consequently for both groups we have case (c) and corepresentations on unitary subgroup are constructed as $\begin{pmatrix} A_1' & 0 \\ 0 & A_2' \end{pmatrix}$

Table 2.2 Double-valued IRs of group C_i

C_i	E	I
A_1'	1	1
A_2'	1	-1

Table 2.3 Double-valued IRs of groups C_2 and C_s

C_2	E	C_2
C_s	E	σ
A_1'	1	i
A_2'	1	$-i$

Example 2.3 Estimate additional degeneracy of spin states due to time-reversal symmetry in a crystal of D_2 symmetry with time reversal, i.e. in Shubnikov group $D_2 + \vartheta D_2$.

Characters of the only doubled-valued IR of group D_2 are presented in Table 2.4. With symmetry D_2, the wavefunctions $p_{1/2}$ and $p_{3/2}$ are transformed as E' and $2E'$, respectively.

When applying Herring criterion (2.2.13) and taking into account that projective factor for all elements except the identity one equal to -1, we have $\sum_R \chi(R^2)\omega(R, R) = 4 - 4 - 4 - 4 = -4$, and also $\nu = -1$.

Thus according to (2.2.13) it is a case (a) and there is no additional degeneracy due to time-reversal symmetry.

Example 2.4 Find additional degeneracy of atomic $p-$ wavefunction in C_{4h} symmetry due to time reversal. To solve this problem the construction of corepresentations of Shubnikov group $M = \mathsf{C}_{4h} + \vartheta \mathsf{C}_{4h}$ is required. In a spinless case, additional multipliers $\nu\left(\vartheta^2\right)$ and $\omega\left(B, B\right)$ in (2.2.12) equal to unity. Since inversion I is an element of C_{4h} it is sufficient to consider characters for pure rotation group C_4 and extend even and odd IRs on the left coset defined by I multiplying by plus or minus sign for gerade and ungerade cases, respectively. The characters of C_{4h} are presented in Table 2.5.

Nonunitary elements for halving subgroup and their squares are presented in Table 2.6. Since all characters of $A_{g(u)}$ and $B_{g(u)}$ are real we obtain that $\sum_{\theta B} \chi(B^2) = |G|$ and corepresentations belong to type (a). Since characters of $E_{1g(u)}$ and $E_{2g(u)}$ for C_{2z} equal to -1 we obtain that $\sum_{\theta B} \chi(B^2) = 0$. Thus, these corepresentations of Shubnikov group $M = C_{4h} + \theta C_{4h}$ belong to type (c) and are constructed as $E'_{1g} + E'_{2g}$ and $E'_{1u} + E'_{2u}$. The $p-$ function with $m = 0$ belongs to IR A_u of C_{4h} and there is no additional degeneracy due to time-reversal symmetry. On the other hand, $p-$functions with $m = 1$ and $m = -1$ belong to IRs E'_{1u} and E'_{2u}, respectively, and they are degenerated due to time-reversal symmetry.

Table 2.4 Character of double-valued IR of group D_2

D_2	E	C_{2x}	C_{2y}	C_{2z}
E'_1	2	0	0	0

Table 2.5 Irreducible representations of C_{4h}

C_{4h}	E	C_{4z}	C_{2z}	C_{4z}^3
$A_{g(u)}$	1	1	1	1
$B_{g(u)}$	1	-1	1	-1
$E'_{1g(u)}$	1	i	-1	$-i$
$E'_{2g(u)}$	1	$-i$	-1	i

Table 2.6 Nonunitary elements and their squares for the left coset ϑC_4

$B = \vartheta R_i$	ϑE	ϑC_{4z}	ϑC_{2z}	ϑC_{4z}^3
B^2	E	C_{2z}	E	C_{2z}

Example 2.5 Find magnetic groups originating from point group $4mm = C_{4v}$ and corepresentations of one of these groups, namely, of $4m'm'$.

It is seen in character table of group C_{4v} (Table 1.3) that there are three nontrivial one-dimensional IRs of group, namely, A_2, B_1 and B_2. Magnetic group induced by these IRs is shown in Fig. 2.3. However, Shubnikov groups arising from IRs B_1 and B_2 (see Fig. 2.3b and c, respectively) differ by permutation of horizontal axes and are

Table 2.7 Relations for antiunitary elements of group $4m'm'$

R	E	C_{4z}	C_{2z}	C_{4z}^3
$B = \vartheta\sigma_y R$	$\vartheta\sigma_y$	$\vartheta\sigma_b$	$\vartheta\sigma_x$	$\vartheta\sigma_a$
B^2	E	E	E	E
BA^{-1}	E	C_{4z}^3	C_{2z}	C_{4z}

Table 2.8 Corepresentations of nonunitary elements of group $4m'm'$. (IRs of a unitary subgroup C_4 are obtained from Table 2.5 by dropping subscripts $g(u)$). The upper (lower) signs correspond to plus (minus) signs in formula (2.2.9)

B	$\theta\sigma_y$	$\theta\sigma_b$	$\theta\sigma_x$	$\theta\sigma_a$
$\Delta(A)$	± 1	± 1	± 1	± 1
$\Delta(B)$	± 1	∓ 1	± 1	∓ 1
$\Delta(E_1)$	± 1	$\mp i$	∓ 1	$\pm i$
$\Delta(E_2)$	± 1	$\pm i$	∓ 1	$\mp i$

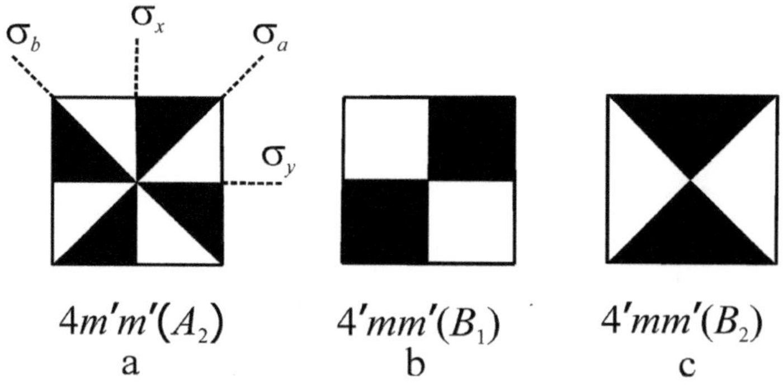

Fig. 2.3 Shubnikov groups arising from C_{4v} point group. Note that groups arising from B_1 and B_2 are equivalent in abstract sense

equivalent in abstract sense. Thus, there are two magnetic groups: $4m'm'$ and $4'mm'$ with unitary subgroups C_4 and C_{2v}, respectively.

Shubnikov group $4m'm'$ corresponds to ferromagnetic orientations of magnetic moments in $z-$direction (see Table 7.7 in [19]). IRs of the unitary subgroup C_4 are obtained from Table 2.5 by dropping subscripts $g(u)$. It is convenient for the problems, considered in Chap. 4 to choose $A = \theta\sigma_y$ as the representative element of antiunitary left coset. Supplementary elements for the construction of corepresentations, namely, antiunitary elements $B = AR$, their squares B^2 and transformed elements BA^{-1} are presented in Table 2.7. It is seen from this table that for all antiunitary elements $B^2 = E$. Thus, we have for the sum in the Herring criterion $\sum_{\theta B \in AG} \chi(B^2) = |G|$ and all corepresentations belong to the type (a) which are constructed as

$$\Delta(R) = D(R), \;\; \Delta(B) = \pm D(BA^{-1})\, N \tag{2.3.1}$$

where $NN^* = +D\left(A^2\right)$. In our case $A^2 = E$ and $N = 1$.

Corepresentations for non-unitary elements are presented in Table 2.8.

Chapter 3
Induced Representation Method in the Theory of Electron Structure

The topic of induced representations of groups is now all the rage among physicists and mathematicians as a result of a sudden general awakening to the extraordinary power and simplicity of this notion.

A.J. Coleman [22]

With a brilliant idea, Frobenius succeeded in constructing representations of a group from those of an arbitrary subgroup

W. Ledermann [23]

Abstract Induced representation method makes it possible to construct representation of the entire group G on the basis of the irreducible representation of its subgroup H and the structure of the left coset decomposition of G with respect to H. The application of theorems of induced representation method, i.e. Frobenius reciprocity theorem and a set of Mackey theorems makes it possible to simplify symmetry analysis of quantum systems with subgroups of symmetry. Following problems of the theory of electron structure are considered making use of induced representation method: (a) Normal vibrations of molecules. A method based on Frobenius reciprocity makes it possible to simplify the existing method. It is shown that the analysis of vibrations can be performed by checking the orthogonality of IRs of the local subgroup and of the whole group without construction transformation matrix on the whole group. (b) MO (molecular orbitals) or SALC (symmetry-adapted linear combinations) in clusters are also analyzed by making use of Frobenius reciprocity theorem. Also the reciprocity theorem is generalized on matrix elements and transitivity of induction theorem is used to label repeating IRs in large symmetrical clusters. Hybrid orbitals in molecules are also considered in a framework of induced representation method. (c) A short theory of space-group IRs based on induced representations is presented. These sections are introductory for Chap. 4. (d) The examples are presented of the application of induced representation method to compatibility relations, selection rules, and SALCs in solids, making use Mackey intertwining number theorem the difference of SALCs in symmorphic and non-symmorphic groups is established.

V. G. Yarzhemsky, *Novel Group Theoretical Methods for Electron Structure Theory*,
https://doi.org/10.1007/978-3-031-72802-0_3

3.1 Systems with Subgroups of Symmetry and Induced Representations

When describing such many-electron systems as molecules or solids, separating the symmetry subgroup H of the whole symmetry group G makes it possible in many cases to classify the wavefunctions by introducing additional quantum numbers on the subgroup. In the theory of space groups, the important subgroup is a group H of the wavevector k (little group) and the label κ of IR (irreducible representation) D^{κ} (small IR) of H is the labels of IRs of the whole space group. In symmetrical molecules, local symmetry groups H of atoms (other than the central atom) are usually used to construct symmetry-adapted wavefunctions and normal vibrations, and the labels of IRs of H are useful physical quantum numbers. For example, the indices σ and π of molecular orbitals are related to IRs of local subgroup. σ-orbitals include $s-$ and p_z-atomic wavefunctions and $\pi-$orbitals include p_x- and p_y-atomic wavefunctions. (Usually local $z-$axis is directed toward the center of symmetry of the molecule.) Description of wavefunctions in systems with symmetry subgroups is most naturally given by the method of induced representations, which is based on the construction of group representations based on the IRs of its subgroup. The induced representation method also makes it possible to classify wavefunctions by the labels of IRs of other physical subgroups, what become especially important in complex quantum systems.

Let H be a subgroup of a finite group G, and $\Phi = \{\varphi_i,\ i = 1, ..q\}$ be a basis set of a $q-$dimensional IR B of H. Let us decompose G into the left cosets relative to H:

$$G = \sum_{\sigma=1}^{n} s_\sigma H,\ n = \frac{|G|}{|H|} \tag{3.1.1}$$

It follows from the decomposition of the group into the left cosets that any element $g \in G$ can be represented as a product of some left coset representative s_ν and an element $h_j \in H$. It should be noted that any element of the given left coset can be chosen as a coset representative. However, if some choice of representatives of the left cosets is made, then it should not change. Acting by the left coset representatives on the initial basis Φ, we obtain a basis set Ω on the whole group:

$$\Omega = \{\omega_{\sigma i} = s_\sigma \varphi_i,\ i = 1, ..q,\ \sigma = 1, ..n\}.. \tag{3.1.2}$$

Let us show that the basis Ω is invariant under the action of the elements of the group G. Consider the action of a group element $g = s_\nu h_j$ on an arbitrary element of the basis Ω:

$$g\omega_{\sigma i} = s_\nu h_j s_\sigma \varphi_i = s_\xi h_k \varphi_i = \sum_p s_\xi B(h_k)_{ip}\varphi_p = \sum_p B(h_k)_{ip}\omega_{\xi p} \tag{3.1.3}$$

The q-dimensional representation $B(h)$ of the subgroup H, which is not required to be irreducible, can be formally extended on the whole group G as follows:

$$B(g) = \left\{ \begin{array}{c} B(g), \text{ if } g \in H \\ 0, \text{ if } g \notin H \end{array} \right\}. \tag{3.1.4}$$

For every element of $g \in G$, we define nq-dimensional matrix $D(g)$:

$$D(g) = \begin{pmatrix} B(s_1^{-1}gs_1) & B\left(s_1^{-1}gs_2\right) & \ldots & B\left(s_1^{-1}gs_n\right) \\ B(s_2^{-1}gs_1) & B\left(s_2^{-1}gs_2\right) & .. & B\left(s_2^{-1}gs_n\right) \\ \ldots & \ldots & \ldots & \ldots \\ B(s_n^{-1}gs_1) & B\left(s_n^{-1}gs_2\right) & \ldots & B\left(s_n^{-1}gs_n\right) \end{pmatrix} \tag{3.1.5}$$

where s_i are the left coset representatives in (3.1.1). Frobenius found that the matrix thus defined implements the representation of the group G, i.e.

$$D(p)D(g) = D(pg),\ p, g \in G \tag{3.1.6}$$

The part of this equation for the (i, j) block is written as

$$\sum_{r=1}^{n} B(s_i^{-1}ps_r)B(s_r^{-1}gs_j) = B(s_i^{-1}pgs_j) \tag{3.1.7}$$

Let's consider two possible cases. In the first case, $s_i^{-1}pgs_j \notin H$ and there must be a zero matrix on the right-hand side of (3.1.7). To do this, for any $(1 \leq r \leq n)$ there must be either $s_i^{-1}ps_r \notin H$ or $s_r^{-1}gs_j \notin H$. In the opposite second case for some r two relations is fulfilled $s_i^{-1}ps_r \in H$ and $s_r^{-1}gs_j \in H$. Consequently we can write that

$$s_i^{-1}ps_rs_r^{-1}gs_j = s_i^{-1}pgs_j \in H \tag{3.1.8}$$

Thus in the second case (i, j)-block of the product matrix is nonzero, and $s_i^{-1}pgs_j \in H$. The element gs_j belongs to one of the left cosets, say:

$$gs_j \in s_kH \tag{3.1.9}$$

Then:

$$s_{k-}^{-1}gs_j \ \in H \tag{3.1.10}$$

Since the left-hand side of (3.1.7) is nonzero, the first factor in the nonzero term must also belong to the subgroup H: $s_i^{-1}ps_r \in H$. Therefore, $r = k$ and only one term $B(s_i^{-1}ps_ks_k^{-1}gs_j) = B(s_i^{-1}pgs_j)$ remains on the left-hand side of (3.1.7).

Thus, we have shown that block matrices given by the formula:

$$[B \uparrow G(g)]_{i\mu,j\nu} = \begin{cases} B(s_i^{-1} g s_j)_{\mu\nu}, \text{ if } s_i^{-1} g s_j \in H \\ 0, \text{ if } s_i^{-1} g s_j \notin H \end{cases} \tag{3.1.11}$$

implement a representation of the group, called the induced representation. The sign $\uparrow$ is used to denote induction. Note that in this formula $B(h)$ can be either a reducible or an irreducible representation of the subgroup H. If $B(h)$ is an irreducible representation, then the induced representation can be both reducible and irreducible.

The induced projective representations are defined by similar way, but the nonzero matrices acquire phase factors, which are products of the elements of factor system and are written as [57]

$$\hat{B}(s_i^{-1} g s_j)\omega\left(s_i^{-1}, g s_j\right)\omega\left(g, s_j\right)/\omega\left(s_i^{-1}, s_i\right) \tag{3.1.12}$$

where ω is a factor system of projective representation $\hat{B}$. It can be seen from the above consideration that the construction of induced basis set Ω is similar to the construction of MO (molecular orbital) and normal vibration of symmetrical molecules. We will show below, how the application of Frobenius reciprocity theorem makes it possible to simplify group-theoretical analysis of MO and normal vibrations.

3.2 Frobenius Reciprocity Theorem

Induction means construction of group representations from subgroup representations. The reverse operation is called subduction. Let $D^q(g)$ be the IR of the group G, then one can always define a representation of the subgroup $H \subset G$ by the formula:

$$\left(D^q \downarrow H\right)(h) = D^q(h), \quad h \in H \tag{3.2.1}$$

Under the subduction to a subgroup, an irreducible representation of a group may remain irreducible or become reducible. The results of induction and subduction of representations are connected by the Frobenius reciprocity theorem. Denote by $f(B^k/D^q \downarrow H)$ the frequency of appearance of IR B^k of the subgroup H in the decomposition of the subduced representation $D^q \downarrow H$. Also denote by $f(D^q/B^k \uparrow G)$ the frequency of appearance of the IR D^q of the group G in the decomposition of the induced representation $B^k \uparrow G$. According to the Frobenius reciprocity theorem, these frequencies are equal to each other:

$$f(D^q/B^k \uparrow G) = f(B^k/D^q \downarrow H) \tag{3.2.2}$$

Note that Frobenius reciprocity theorem is also valid for induced projective representation, defined by formula (3.1.12) [57]. To obtain the frequency in the left-hand side of (3.2.2) the characters' summation over all elements of group G is required.

But the sum in the right-hand side (3.2.2) includes summation of known characters on a subgroup H and we can write that

$$f(D^q/B^k \uparrow G) = \frac{1}{|H|}\sum_{h\in H} \chi^*\left[B^k(h)\right]\chi\left[D^q(h)\right] \tag{3.2.3}$$

This remarkable theorem finds its application both in physics and in mathematics. In mathematics, by constructing induction of known representations of a subgroup, representations of a larger group may be found. In physics, induced representations correspond to subgroups that have a well-defined physical meaning. To decompose such states into irreducible components, it is also convenient to use the Frobenius theorem, which, without constructing a complete basis for such states, finds its decomposition into irreducible components based on known representations of the group and subgroup.

The reciprocity theorem is used for the symmetry description of systems with physically distinguished symmetry subgroups. In the theory of electronic structure, these are the problems of analyzing normal vibrations of molecules and construction of symmetry-adapted functions of multicenter systems (molecular orbitals) in terms of linear combinations of atomic orbitals (a so-called MO LCAO method).

3.3 Finding the Symmetry of Normal Vibrations of Molecules

A molecule consisting of n atoms has $3n$ degrees of freedom. Three degrees of freedom correspond to translational motion and three degrees of freedom to rotational motion. Thus, there are $3n - 6$ vibrational degrees of freedom. If the molecule is linear, then the number of rotational degrees of freedom is two and there is one vibrational degree of freedom.

Induced representation method makes it possible to obtain symmetry of vibrations for each orbit of atoms separately. Let G be the symmetry group of the molecule, and let H be the symmetry group of the atom A_1, then the number n of atoms of type A is

$$n = |G|\,/\,|H| \tag{3.3.1}$$

The displacements of the atom A_1 from the equilibrium position are the components of the vector (x_1, y_1, z_1). Under the action of the element of the subgroup H, these coordinates of vector are transformed according to the representation D of this subgroup. The representation D is decomposed into IRs D^k of subgroup H. Displacements of different atoms forming one orbit are not independent. The set of $3n$ displacements of atoms forming one orbit must be transformed according to the representation (reducible) of the entire group G. Let the atom A_1 be mapped to the atom A_i by the left coset representative s_i, and the basis (x_1, y_1, z_1) is mapped to the basis (x_i, y_i, z_i) by the same element s_i. The elements s_i are the left coset representatives in the decomposition of G with respect to H. Then, under the action of the elements of the symmetry group $H_i = s_i H s_i^{-1}$ of the atom A_i, the displacements (x_i, y_i, z_i) are transformed by the representation ${}_iD^k(\bar{h}) = D^k(s_i^{-1}\bar{h}s_i)$ of this group $(\bar{h} \in H_i)$. Thus, we have shown that under the action of the elements of local groups H_i and under the action of left coset representatives s_i, the displacements of atoms are transformed as the basis of the induced representation. Therefore, the complete matrix of coordinate transformations for all atoms of a given orbit coincides with the induced representation $D^k \uparrow G$. To find the symmetry of all normal vibrations, there is no need to build this matrix. It is sufficient to decompose each one of induced representations $D^k \uparrow G$, on IRs Γ^q of G making use of the Frobenius reciprocity theorem (3.2.3).

Such an analysis must be carried out for all orbits, and then the IRs of the group G corresponding to the translational and rotational motion of the entire molecule should be subtracted. The IRs, corresponding to rotations and disagreements, are usually noted in character tables [11]. Thus making use of induced representation method we need only the characters' tables of the local subgroup and of the whole group G, and the construction of total transformation matrix is not required. This technique is illustrated by the following three examples.

Example 3.1 Find symmetry of vibrations of CH_4 molecule of T_d symmetry.

The basis set of atomic displacements in T_d symmetry is shown in Fig. 3.1. The first atom H_1 is in $z-$axis and directions of the coordinate axes of H_1 are the same as that on central atom C. Local symmetry group of atom H_1 is C_{3v}. The basis sets on the other atoms H_i are obtained by C_{2i} rotations about axes passing through the central atom and the middle points of the segments H_1-H_i. It is clear that three rotations C_{2i} together with an identity element implement the left coset decomposition of group T_d of 24 elements with respect to its subgroup C_{3v} of 6 elements. The stretching vibrations (displacement along $z-$axis) of atom H_1 belong to IR A_1 of C_{3v} and the bending vibrations ($x-$ and $y-$displacements) belong to IR E. The basis sets on other atoms are obtained by acting the left coset representative on the basis set of the first atom. It is clear that the procedure of construction of it is plural diaplacements of atoms is the same as that of construction of induced representation basis (3.1.2). Thus basis sets of two vibration modes belong to induced representations $A_1 \uparrow T_d$ and $E \uparrow T_d$. The decomposition of the induced representations can be easily done by Frobenius reciprocity theorem (3.2.3) and making use of characters of C_{3v} and characters of T_d on C_{3v}, presented in Table 3.1. Making use of formula (3.2.3) one obtains that $A_1 \uparrow T_d = A_1 + T_2$ and $E \uparrow T_d = E + T_1 + T_2$. To obtain all vibration modes, one should add displacements of central atom, which belong to IR T_2, and subtract the representation of the whole molecule displacements T_2 and rotations T_1. Thus, one obtains total set of vibration modes of CH_4 molecule:

$A_1 + E + 2T_2$

Thus, in contrast to the standard technique [14, 21], where in the case of CH_4 the character of the 15-dimensional matrix is built and then decomposed by summation

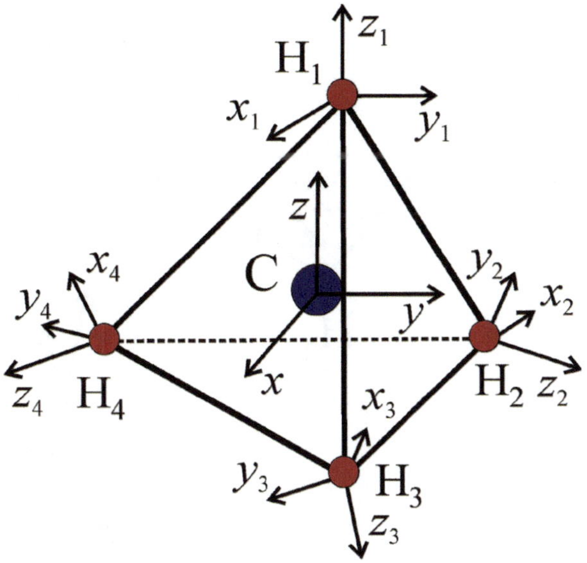

Fig. 3.1 Coordinate systems for displacements of atoms in CH_4 molecule of T_d symmetry

Table 3.1 Character table of group T_d on its subgroup C_{3v} and part of character table of group C_{3v}

Group	IR	$E(1)$	$C_3(2)$	$\sigma_v(3)$
T_d	A_1	1	1	1
	A_2	1	1	−1
	E	2	−1	0
	T_1	3	0	−1
	T_2	3	0	1
C_{3v}	A_1	1	1	1
	E	2	−1	0

over the entire group, we obtained the same result by checking the orthogonality of the tabular characters on the subgroup.

Example 3.2 Find the symmetry of the normal vibrations of the NH_3 molecule.

In the NH_3 molecule, the H atoms form a regular triangle, the N atom is beyond the plane, passing through the three H atoms (see Fig. 3.2). The NH_3 molecule has C_{3v} symmetry and consists of two orbits. The local symmetry of the N atom is C_{3v} and it forms an orbit itself. The coordinates of the displacement vector of the atom N are transformed according to the IRs $A_1 + E$ of the group C_{3v}. Local symmetry of H atoms is C_s (C_s consists of an identity element and a reflection). Let's direct z_1−axis of the initial atom H_1 toward the central atom, y_1−axis in the reflection

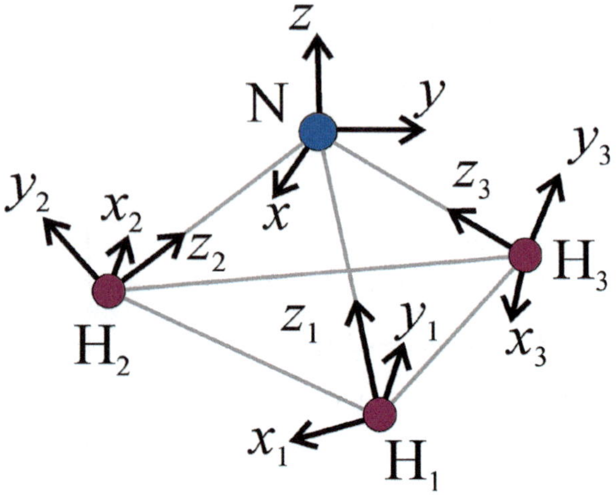

Fig. 3.2 Coordinate systems for displacements of atoms in NH_3 molecule of C_{3v} symmetry

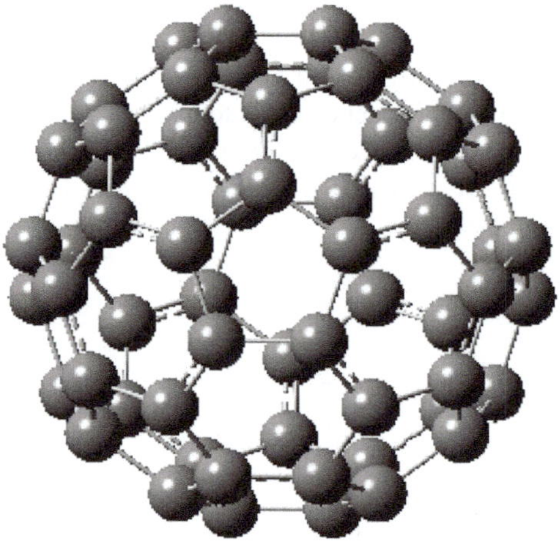

Fig. 3.3 Fullerene C_{60} molecule

plane normal to z_1, and the x_1-axis perpendicular to this plane. In this case, y_1 and z_1 are transformed by IR A, and x_1 is transformed by A' of C_s. The basis sets on other H atoms are obtained from the basis on the first atom by rotations C_{3z} and C_{3z}^2. Induction from C_s into the group C_{3v} results in $2\,(A \uparrow C_{3v}) = 2\,(A_1 + E)$ and $A' \uparrow C_{3v} = A_2 + E$ and the sum is $2A_1 + A_2 + 3E$. Also we should add vector representation $A_1 + E$ for the atom N, subtract $A_1 + E$ (translational motion) and $A_2 + E$- (rotational motion). As a result, we obtain that the set of normal vibrations for the NH_3 molecule includes IRs: $2A_1 + 2E$.

Example 3.3 Find normal vibrations of fullerene C_{60}.

In the C_{60} fullerene, the C atoms are on symmetry planes, outside the symmetry axes (see Fig. 3.3) and local symmetry group of atom C is C_s.

We choose a local coordinate system so that the $z-$axis is in symmetry plane and is directed from the center of symmetry, the x-axis lies in the symmetry plane, and the $y-$axis is perpendicular to the symmetry plane (see Fig. 3.4).

The characters of the group I_h on its subgroup C_s are given in Table 3.3. When using this data and characters of C_s, presented in two last lines of Table 3.2 and Frobenius reciprocity theorem (3.2.3), one obtains that

$A \uparrow I_h = A_g + T_{1g} + T_{2g} + 2G_g + 3H_g + 2T_{1u} + 2T_{2u} + 2G_u + 2H_u$ and

$A' \uparrow I_h = 2T_{1g} + 2T_{2g} + 2G_g + 2H_g + A_u + T_{1u} + T_{2u} + 2G_u + 3H_u$

The complete basis of displacements is transformed by the sum of the induced representations $2A_1 \uparrow I_h + A' \uparrow I_h$. When subtracting IR T_{1u} of translations and IR T_{1g} of rotations, we obtain total set of normal vibrations of C_{60}:

$2A_g + A_u + 3T_{1g} + 4T_{1u} + 4T_{2g} + +5T_{2u} + 6G_g + 6G_u + 8H_g + 7H_u$

From the solution of these three examples, it is clear that, unlike standard methods [1, 14, 15, 21], where the character of the matrix of all displacements of the molecule is built and then decomposed to find normal vibrations, the method of induced rep-

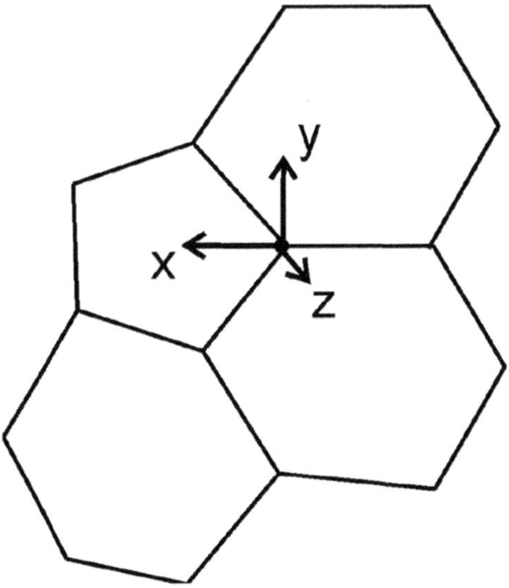

Fig. 3.4 Basis set for vibrations of one atom of C_{60}

Table 3.2 Characters of group C_{3v} on its subgroup C_s and characters of group C_s

Group		E	σ_v
C_{3v}	A_1	1	1
	A_2	1	−1
	E	2	0
C_s	A	1	1
	A'	1	−1

Table 3.3 Characters of group I_h on its subgroup C_s and characters of C_s

	IR	E	$\sigma_{g(u)}$
I_h	$A_{g(u)}$	1	1(−1)
	$T_{1g(u)}$	3	−1(1)
	$T_{2g(u)}$	3	−1(1)
	$G_{g(u)}$	4	0
	$H_{g(u)}$	5	1(−1)
C_s	A	1	1
	A'	1	−1

resentations allows one to obtain identical results from the analysis of the character tables of the entire group and local subgroup only.

3.4 Molecular Orbitals' Symmetry Analysis

In the theoretical consideration of wavefunctions and interactions in molecules and clusters, the MO LCAO approach is used, in which the molecular orbital (the wavefunction of a molecule) is expressed as a linear combination of atomic orbitals. These combinations are constructed for atoms belonging to the same orbits, i.e. group of atoms, connected by symmetry operations and linear combinations of orbitals belonging to certain IR are constructed. Such a linear combination $\Omega(A, l, n, D^k, \Gamma^q)$ is characterized by the atom type A, principal n and orbital l quantum numbers, the IR of the local group D^k and the IR of the entire group Γ^q. Since the Hamiltonian is totally symmetric in the symmetry group G, only the wavefunctions that belong to the same IR of the entire group can interact. The wavefunctions of a single orbit are constructed by methods of group theory. From the wavefunctions of different orbits, but belonging to the same IR, using the self-consistent field methods, linear combinations are constructed that minimize the total energy. We symbolically write such a function as

$$\Psi_0(\Gamma^q) = \alpha\Omega(A_1, l, n, D^k, \Gamma^q) + \beta\Phi(A_2, l', n', B^j, \Gamma^q) + \gamma\Theta(A_3, l'', n'', C^p, \Gamma^q) \quad (3.4.1)$$

The combinations Ω, Φ and Θ, which are also called symmetry-adapted linear combinations of atomic functions (SALC), are constructed group-theoretically. It will be shown below, how the induced representation method makes it possible to simplify this procedure.

Consider a p-orbitals in octahedral complex, in which six atoms are in coordinate directions (see Fig. 3.5), where short notations x, y and z are used instead of p_x, p_y and p_z. Directions of the axes of the first center in the six-atomic orbit are the

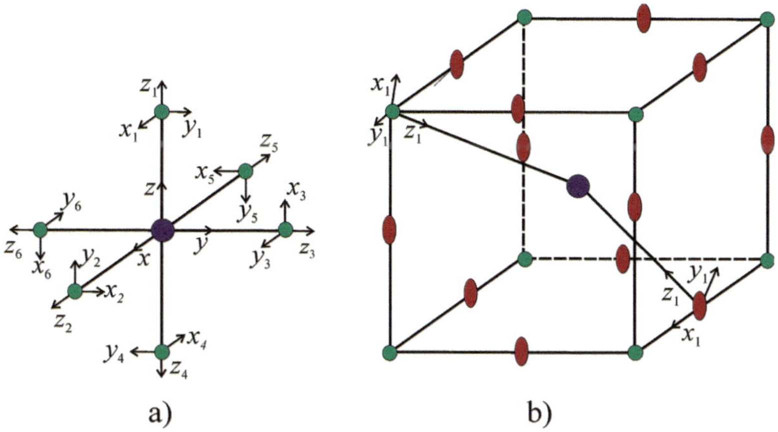

Fig. 3.5 Three orbits of atoms in O_h symmetry **a** octahedron, **b** cube (circles) and cuboctahedron (ellipses)

same as that of central atom. The axes of the second and third centers are obtained by counterclockwise rotation about the axis (111) at angles $2\pi/3$ and $4\pi/3$, respectively, and the rest of basis set is obtained by the space inversion. To convert basis functions on the centers 4, 5 and 6 into standard form [9] one can rename x and y basis functions, but it is not essential in our general consideration. Functions z_1 and (x_1, y_1) belong to IRs A_1 and E, respectively, of local group C_{4v} of the first atom. These orbitals are called σ- and π-orbitals, respectively. Under the action of the elements of O_h group functions (z_i, i=1..6) and (x_i, y_i, i=1,..6) are transformed independently by induced representations $A_1 \uparrow O_h$ and $E \uparrow O_h$ of dimensions 6 and 12, respectively. Group-theoretical analysis of molecular orbitals can be done by Frobenius reciprocity theorem similar to the analysis of normal vibrations by making use of formula (3.2.3). Characters of IRs of O_h and of its subgroup C_{4v}, C_{3v} and C_{2v} are presented in Table 3.4. When applying formula (3.2.3), i.e. checking the orthogonality of characters of group and subgroup on the subgroup we obtain for σ-orbitals, i.e. originating from A_1:

$$A_{1(\mathsf{C}_{4v})} \uparrow G = A_{1g} + E_g + T_{1u} \tag{3.4.2}$$

Here and wherever additional explanation is required, we write in the form of a subscript the subgroup from which the induction is carried out. Since in C_{4v} symmetry $s-$orbitals also belong to IR A_1, there is another set of $\sigma-$orbitals A_{1g}, E_g and T_{1u}, originating from $s-$orbitals. For π-orbitals, i.e. originating from IR E we obtain from Table 3.4:

$$E_{(\mathsf{C}_{4v})} \uparrow G = T_{1g} + T_{1u} + T_{2g} + T_{2u} \tag{3.4.3}$$

Suppose that there are $s-$, $p-$ and $d-$orbitals on the central atom. The $s-$ and $p-$orbitals of central atom are unsplit and belong to IRs A_{1g} and T_{1u}, respectively,

Table 3.4 Characters of group O_h on its subgroups C_{4v}, C_{3v} and C_{2v} with the characters of groups C_{4v}, C_{3v} and C_{2v}

O_h (C_{4v})	E	C_{2z}	$2C_{4z}$	$2\sigma_v$	$2\sigma_{v'}$	O_h (C_{3v})	E	$2C_3$	$3\sigma_v$	O_h (C_{2v})	E	C_2	σ_v	$\sigma_{v'}$
$A_{1g}(A_{1u})$	1	1	1	1(−1)	1(−1)	$A_{1g}(A_{1u})$	1	1	1(−1)	$A_{1g}(A_{1u})$	1	1	1(−1)	1(−1)
$A_{2g}(A_{2u})$	1	1	−1	1(−1)	−1(1)	$A_{2g}(A_{2u})$	1	1	−1(+1)	$A_{2g}(A_{2u})$	1	−1	1(−1)	−1(1)
$E_g(E_u)$	2	2	0	2(−2)	0	$E_g(E_u)$	2	−1	0	$E_g(E_u)$	2	0	2(−2)	0
$T_{1g}(T_{1u})$	3	−1	1	−1(1)	−1(1)	$T_{1g}(T_{1u})$	3	0	−1(+1)	$T_{1g}(T_{1u})$	3	−1	−1(1)	−1(1)
$T_{2g}(T_{2u})$	3	−1	−1	−1(1)	1(−1)	$T_{2g}(T_{2u})$	3	0	1(−1)	$T_{2g}(T_{2u})$	3	1	−1(1)	1(−1)
C_{4v}						C_{3v}				C_{2v}				
A_1	1	1	1	1	1	A_1	1	1	1	A_1	1	1	1	1
A_2	1	1	1	−1	−1	A_2	1	1	−1	B_1	1	−1	1	−1
B_1	1	1	−1	1	−1	E	2	−1	0	B_2	1	−1	−1	1
B_1	1	1	−1	−1	1									
E	2	−2	0	0	0									

$d-$orbitals split into two states $E_g(2z^2-x^2-y^2, x^2-y^2)$ and $T_{2g}(yz, xz, xy)$. Hence, it follows that $A_{1g}(s)$, $T_{1u}(p)$ and $E_{2g}(d)$ orbitals of the central atom form bonds with σ-orbitals of ligands. As to $\pi-$orbitals, T_{1u} and T_{2g} interact with $p-$ and $d-$orbitals of the central atom, but the orbitals T_{1g} and T_{2u} don't interact with the $s-$, $p-$ and $d-$orbitals of central atom. Since it is usually assumed that $\sigma-$interaction is stronger than $\pi-$interaction, we can expect that the energy of $E_g(d)-$electrons is lower than that of $T_{2g}(d)$-electrons.

Similar analysis can be easily performed for cubic eightfold coordination in O_h symmetry shown in Fig. 3.5b. In this case, local p_z-orbitals and $(p_x, p_y)-$orbitals belong to IRs A_1 and E of local symmetry group C_{3v}, respectively. Making use of characters of group O_h on C_{3v} subgroup and characters of C_{3v} and applying Frobenius reciprocity theorem (3.2.3) one easily obtains possible symmetries of $\sigma-$ and $\pi-$orbitals, respectively:

$$A_{1(C_{3v})} \uparrow O_h = A_{1g} + A_{2u} + T_{1u} + T_{2g} \tag{3.4.4}$$

$$E_{(C_{3v})} \uparrow O_h = E_g + E_u + T_{1g} + T_{1u} + T_{2g} + T_{2u}$$

If an atom is placed in the middle of the edge of the cube, where the local symmetry is C_{2v}, then acting by the left cost representatives, we obtain a cuboctahedral orbit consisting of 12 atoms (see Fig. 3.5b). The local $z-$axis is directed to the central atom and belongs to IR A_1 of C_{2v}. The $\sigma-$orbitals belong to the induced representation $A_{1(C_{2v})} \uparrow O_h$, which can be easily expanded making use of characters of group and subgroup presented in Table 3.4 and formula (3.2.3), and we obtain

$$A_{1(C_{2v})} \uparrow O_h = A_{1g} + E_g + T_{1u} + T_{2g} + T_{2u} \tag{3.4.5}$$

It is seen from Fig. 3.5b that reflection in coordinate plane doesn't change the sign of local $y-$function, but the reflections in diagonal plane change its sign and it belongs to IR B_1 (see Table 3.4). Also it can be verified that x belongs to IR B_2. Thus we obtain complete set of $\pi-$orbitals in cuboctahedral coordination:

$$B_{1(C_{2v})} \uparrow O_h = A_{2g} + E_g + T_{1g} + T_{1u} + T_{2u}$$

$$B_{2(C_{2v})} \uparrow O_h = A_{2u} + E_u + T_{1g} + T_{1u} + T_{2g}$$

Since atomic $p-$orbitals and atomic displacements are represented by the same way, i.e. by function x, y and z, we can use Fig. 3.1 to illustrate $p-$electron MO basis set in T_d symmetry. Thus, we obtain by the similar way MO of four atomic orbit in T_d symmetry:

$$A_{1(C_{3v})} \uparrow T_d = A_1 + T_2 \tag{3.4.6}$$

$$E_{(C_{3v})} \uparrow T_d = E + T_1 + T_2 \tag{3.4.7}$$

The first set of IRs corresponds to $\sigma-$orbitals and the second set to $\pi-$orbitals. The orbitals of the central atom in T_d symmetry are as follows: $A_1(s)$, $T_2(p_x, p_y, p_z)$, $E(d_{2z^2-x^2-y^2}, d_{x^2-y^2})$, and $T_2(d_{yz}, d_{xz}, d_{xy})$. Comparing the symmetries of orbitals of central atom and of ligands, we see that the orbitals $T_2(d_{yz}, d_{xz}, d_{xy})$ can form bonds with $\sigma-$ and $\pi-$orbitals of ligands, but $E(d_{2z^2-x^2-y^2}, d_{x^2-y^2})$ orbitals of the central atom can form bonds with $\pi-$orbitals of ligands only. Thus we can expect that in T_d symmetry the T_2 orbital lies lower in energy than the E- orbital.

3.5 Induced Representations and Hybrid Orbitals

In the molecular orbital theory, the interaction of orbitals of the central atom with orbitals constructed from orbitals of ligands is considered. In this approach, orbitals belonging to different IRs have different energies and do not interact with each other. It was also found to be useful to consider chemical bonds, i.e. electron densities between central atom and ligands [59]. The $s-$ and p_z-orbitals of ligands were already directed to the central atom (p_z-orbitals due to the choice of local coordinate systems). The problem is to construct from all the orbitals of the central atom electron densities directed to ligands. To do this, consider first the solution of the inverse problem. Consider bonds b_i $(i = 1..n)$ directed from the central atom to equivalent ligands. If the symmetry group of molecule is G and symmetry group of equivalent ligands is H, then the dimension n of this basis set is equal to $|G| / |H|$. Such bonds form the basis of the induced representation $A_{1(H)} \uparrow G$. Making use of Frobenius reciprocity theorem this induced representation can be easily decomposed in terms of the IRs D^q of the point group G:

$$A_{1(H)} \uparrow G = \sum_q f_q D^q \tag{3.5.1}$$

In all cases under consideration, the frequencies f_q do not exceed unity and we omit additional quantum numbers to denote repeating IRs D^q.

The basis functions μ^q_α $(\alpha = 1, .. |D^q|)$ of each point group IR D^q in formula (3.5.1) may be expressed as linear combination of "bonds" b_i:

$$\mu^q_\alpha = \sum_{i=1..n} C^i_{q,\alpha} b_i \tag{3.5.2}$$

The inverse matrix transforms basis functions μ^q_α of IR D^q into bonds:

$$b_i = \sum_{q,\alpha} (C^{-1})^i_{q,\alpha} \mu^q_\alpha \tag{3.5.3}$$

Up to this point, the functions μ_{α}^{q} were expressed in terms of "bond" and had no exact physical meaning. However, if now we substitute the basis functions of the IRs D^q of the point group expressed in terms of the atomic wavefunctions into formula (3.5.3) as μ_{α}^{q}, then we obtain the bond basis b_i expressed in terms of atomic orbitals. Such a basis set of equivalent orbitals expressed through orbital central atoms is called hybrid orbitals .

The induced representation $A_{1(H)} \uparrow G$ for T_d and O_h was considered in previous section, and we can use these results to obtain hybrid basis sets. It is seen from formula (3.4.6) that in T_d symmetry such induced representation is decomposed into two IRs $A_1 + T_2$. Thus we obtain two possible hybrid basis sets for a tetrahedral molecule:

$$s + \left\{p_x, p_y, p_z\right\}, s + \left\{d_{xy}, d_{yz}, d_{xy}\right\} \text{ or } f_{xyz} + \left\{f_{x(y^2-z^2)}, f_{y(z^2-x^2)}, f_{z(x^2-y^2)}\right\} \tag{3.5.4}$$

The first set sp^3 is related to carbon atom in CH_4 molecule and the second sd^3 corresponds to the complexes of $d-$elements in crystals. For the sixfold coordination in O_h symmetry (3.4.2) $A_{1(C_{4v})} \uparrow G$ is decomposed into $A_{1g} + T_{1u} + E_g$, and the hybrid basis set is written as

$$s + \left\{p_x, p_y, p_z\right\} + \left\{d_{2z^2-x^2-y^2}, d_{x^2-y^2}\right\} \tag{3.5.5}$$

For eightfold cubic coordination $A_{1(\mathrm{C}_{3v})} \uparrow G$ is decomposed into $A_{1g} + T_{2g} + A_{2u} + T_{1u}$ (3.4.4) and the full atomic hybrid basis set is written as

$$s + \left\{d_{xy}, d_{yz}, d_{xy}\right\} + \left\{f_{xyz}\right\} + \left\{p_x, p_y, p_z\right\} \tag{3.5.6}$$

In this coordination, the f_{xyz}-orbitals are required to form hybrid bonds. Also instead of $p-$ basis set of T_{1u} we can substitute the basis set constructed from $f-$orbitals the same IR $\left\{f_{x^3}, f_{y^3}, f_{z^3}\right\}$. Thus eightfold cubic coordination is favorable for $f-$elements.

For 12-fold cuboctahedra coordination (3.4.5) $A_{1(C_{2v})} \uparrow G$ is decomposed into $A_{1g} + T_{1u} + E_g + +T_{2g} + T_{2u}$ and the full atomic hybrid basis set is written as

$$\begin{aligned} s &+ \left\{p_x, p_y, p_z\right\} + \left\{d_{2z^2-x^2-y^2}, d_{x^2-y^2}\right\} + \left\{d_{xy}, d_{yz}, d_{xy}\right\} \\ &+ \left\{f_{x(y^2-z^2)}, f_{y(z^2-x^2)}, f_{z(x^2-y^2)}\right\} \end{aligned} \tag{3.5.7}$$

Example 3.4 Hybrid orbitals for icosahedral 12-fold coordination (see Fig. 1.4). In this case, local symmetry of atoms in the shell is C_{5v}. The part of character table of I_h group is presented in Table 3.5 together with characters of C_{5v} group. It is seen in Table 3.5 that totally symmetric A_1 IR of C_{5v} is not orthogonal to IRs A_g, T_{1u}, T_{2u} and H_g of I_h. It is seen from Table 1.11 that these IRs correspond to $s-$, $p-$, $d-$ and $f-$orbitals, respectively, and hybrid can be written as $sp^3d^5f^3$.

Table 3.5 Character tables of C_{5v} and I_h groups on the group C_{5v}

I_h	$1E$	$2C_5^{1,4}$	$2C_5^{2,3}$	$5IC_2^1$
A_g	1	1	1	1
A_u	1	1	1	−1
T_{1g}	3	$\frac{1+\sqrt{5}}{2}$	$\frac{1-\sqrt{5}}{2}$	−1
T_{1u}	3	$\frac{1+\sqrt{5}}{2}$	$\frac{1-\sqrt{5}}{2}$	1
T_{2g}	3	$\frac{1-\sqrt{5}}{2}$	$\frac{1+\sqrt{5}}{2}$	−1
T_{2u}	3	$\frac{1-\sqrt{5}}{2}$	$\frac{1+\sqrt{5}}{2}$	+1
G_g	4	−1	−1	0
G_u	4	−1	−1	0
H_g	5	0	0	1
H_u	5	0	0	−1
C_{5v}				
A_1	1	1	1	1
A_2	1	1	1	−1
E_1	2	$\frac{-1+\sqrt{5}}{2}$	$\frac{-1-\sqrt{5}}{2}$	0
E_2	2	$\frac{-1-\sqrt{5}}{2}$	$\frac{-1+\sqrt{5}}{2}$	0

3.6 Projection Operators in Multicenter Systems

The induced representation basis $\Phi = \{\phi_{\sigma i} = s_\sigma \varphi_i,\ i = 1, ..k,\ \sigma = 1, ..n\}$ which is obtained by the action of left coset representatives s_σ (3.1.1) on the subgroup basis sets φ_i is reducible, and may be decomposed on reducible basis sets belonging to IRs Γ^q of the whole group G by standard projection operator method by summation over the entire group. However, the application of induced representation method makes possible to simplify this procedure. According to generalized Frobenius reciprocity theorem [18], it is sufficient to obtain the reduction coefficients of IR Γ^q of the whole group G on IR D^k of the local group H, and then to transform these coefficients by the known matrices $\Gamma^q(s_\sigma)$, where s_σ are representatives in the left coset decomposition of G with respect to H.

Let there be a system of equivalent centers with full symmetry group G and local symmetry group H. Let a basis $\{\varphi_i,\ i = 1, ...k\}$ of k−dimensional IR D^k of the group H be given on the first center. Let us fix the coordinate system (x_1, y_1, z_1) on the first center and coordinate systems on other centers $(x_\sigma, y_\alpha, z_\sigma)$ are obtained by the action of left coset representatives s_σ on the coordinate system of the first center. We choose the wavefunction bases on the other centers in a similar way:

Our task is to construct orbitals ψ_α^q, transforming by the α row of the IR Γ^q of symmetry group G of the molecule in the induced basis $\phi_{\sigma i}$:

$$\psi_\alpha^q = \sum_{i,\sigma} C_{q,\alpha}^{k\sigma i} \phi_{\sigma i} \tag{3.6.1}$$

The method of finding the reduction coefficients $C_{q,\alpha}^{k\sigma i}$ of induced representation is similar to the standard method of finding the Clebsch-Gordan coefficients for finite groups, and the equation is written as

$$C_{q\alpha}^{k\lambda i} C_{q\beta}^{k\nu j*} = \frac{|\Gamma^q|}{|G|} \sum_g \left(D^k \uparrow G\right)_{\lambda i, \nu j} \Gamma_{\alpha\beta}^q \tag{3.6.2}$$

where $|\Gamma^q|$is the dimension of Γ^q and the sum runs over the whole group. First, we calculate the square of the reduction coefficients for the first atom.

The elements of the block (1, 1) of the induced representation are nonzero only if $g \in H$, then we have for $i = j$:

$$C_{q\alpha}^{k1i} C_{q\alpha}^{k1i*} = \frac{|\Gamma^q|}{|G|} \sum_h D_{ii}^k(h) \Gamma_{\alpha\alpha}^{q*}(h) \tag{3.6.3}$$

The sum on the right-hand side is equal, up to a constant factor, to the square of the corresponding reduction coefficient of the IR Γ^q on the subgroup H. It should be noted that in the case of reduction coefficients, as in the case of Clebsch-Gordan coefficients, the phase of the first coefficient is arbitrary. It follows from this formula that for the first atomic center all reduction coefficients are equal (up to a normalization factor) to the coefficients $\bar{C}_{ki}^{q\alpha}$ of the reduction of IR Γ^q when group G is reduced to its subgroup H. Thus for the first block, we have

$$C_{q\alpha}^{k1i} = \sqrt{\frac{|\Gamma^q|\,|H|}{|G|\,|D^k|}} \bar{C}_{ki}^{q\alpha} \tag{3.6.4}$$

To obtain all the reduction coefficients, it suffices to consider the elements of the first columns of matrices $D^k \uparrow G$ and Γ^q. Consider a σ-block row (in the first block column). According to the definition of the induced representation, the $(\sigma, 1)$ block of the induced representation is nonzero only for elements g belonging to the left coset $s_\sigma H$. Then, we have

$$C_{q\alpha}^{k\lambda i} C_{q1}^{k11*} = \frac{|\Gamma^q|}{|G|} \sum_h D^k(h)_{i1} \Gamma_{\alpha 1}^q(s_\sigma h) \tag{3.6.5}$$

Given that

$$\Gamma_{\alpha 1}^q(s_\sigma h) = \sum_\beta \Gamma_{\alpha\beta}^q(s_\sigma) \Gamma_{\beta 1}^q(h) \tag{3.6.6}$$

we get

$$C_{q\alpha}^{k\lambda i} = \sum_\beta \Gamma_{\alpha\beta}^q(s_\sigma) C_{q\alpha}^{k1i} \tag{3.6.7}$$

Consequently, the reduction coefficients for σ−block (σ−atomic center) are obtained by transforming the reduction coefficients for the first block $C_{q\alpha}^{k1i}$ (the first atomic center) by the matrix of IR of the whole group for the left coset representative $\Gamma^q(s_\sigma)$.

Thus, to find symmetrized combinations for the orbit of a atoms, one must

1. Find the coefficients of projection of the basis functions of the first atom A_1 onto the rows of IR $\Gamma_{\alpha\beta}^q$.

2. The coefficients for the remaining atoms A_σ are found by transforming the obtained row of coefficients by the matrix of representatives of the left cosets $\Gamma_{\alpha\beta}^q(s_\sigma)$.

Example 3.5 For O_h symmetry and sixfold coordination build from six s−orbitals (Fig. 3.5a), symmetrized combinations of A_g, E_g and T_{1u} symmetry.

(a) A_{1g} orbital. The reduction coefficient (3.6.4) of IR A_{1g}, when $O_h \downarrow C_{4v}$ equals to $1/\sqrt{6}$. Transformation matrix $\Gamma_{\alpha\beta}^q$ in (3.6.7) is one-dimensional identity matrix, and we obtain the following linear combinations:

$$A_g : (s_1 + s_2 + s_3 + s_4 + s_5 + s_6)/\sqrt{6} \tag{3.6.8}$$

(b) E_g orbital. The matrices of IR E_g of O_h group for the elements of subgroup C_{4v} are as follows:

$$\underset{E}{\begin{pmatrix} 1 & 0 \\ 0 & 1 \end{pmatrix}} \underset{C_2^z}{\begin{pmatrix} 1 & 0 \\ 0 & 1 \end{pmatrix}} \underset{C_{4z}}{\begin{pmatrix} -1 & 0 \\ 0 & 1 \end{pmatrix}} \underset{C_{4x}^3}{\begin{pmatrix} -1 & 0 \\ 0 & 1 \end{pmatrix}}$$

$$\underset{\sigma_x}{\begin{pmatrix} 1 & 0 \\ 0 & 1 \end{pmatrix}} \underset{\sigma_y}{\begin{pmatrix} 1 & 0 \\ 0 & 1 \end{pmatrix}} \underset{\sigma_a}{\begin{pmatrix} -1 & 0 \\ 0 & 1 \end{pmatrix}} \underset{\sigma_b}{\begin{pmatrix} -1 & 0 \\ 0 & 1 \end{pmatrix}}$$

Also the following matrices of E_g for the left coset representatives transforming s_1 into s_2 and s_3, respectively, will be used: $\underset{C_3}{\begin{pmatrix} -\frac{1}{2} & \frac{\sqrt{3}}{2} \\ \frac{-\sqrt{3}}{2} & -\frac{1}{2} \end{pmatrix}} \underset{C_3^2}{\begin{pmatrix} -\frac{1}{2} & \frac{-\sqrt{3}}{2} \\ \frac{\sqrt{3}}{2} & -\frac{1}{2} \end{pmatrix}}$

Since the s−orbital is completely symmetric we easily obtain the projection coefficients for the rows of E_g:

$$\frac{1}{8\sqrt{3}} \sum_{h \in C_{4v}} \Gamma_{11}^q = 0 \text{ and } \frac{1}{8\sqrt{3}} \sum_{h \in C_{4v}} \Gamma_{22}^q = 1/\sqrt{3} \tag{3.6.9}$$

The coefficients of projecting the orbitals s_2 and s_3 are found by transforming of the obtained column of coefficients by the IRs matrices Γ^q for the rotations C_3 and C_3^2 about (111) axis:

$$\underset{C_3}{\begin{pmatrix} -\frac{1}{2} & \frac{\sqrt{3}}{2} \\ \frac{-\sqrt{3}}{2} & -\frac{1}{2} \end{pmatrix}} \begin{pmatrix} 0 \\ 1/\sqrt{3} \end{pmatrix} = \begin{pmatrix} \frac{1}{2} \\ -\frac{1}{2\sqrt{3}} \end{pmatrix} \tag{3.6.10}$$

$$\overset{C_3^2}{\begin{pmatrix} -\frac{1}{2} & \frac{-\sqrt{3}}{2} \\ \frac{\sqrt{3}}{2} & -\frac{1}{2} \end{pmatrix}} \begin{pmatrix} 0 \\ 1/\sqrt{3} \end{pmatrix} = \begin{pmatrix} \frac{-1}{2} \\ -\frac{1}{2\sqrt{3}} \end{pmatrix} \tag{3.6.11}$$

Due to the parity of the IR E_g the coefficients for atoms connected by inversion are the same and we obtain finally

$$E_g : \begin{pmatrix} \frac{1}{2}(s_2 - s_3 + s_5 - s_6) \\ \frac{1}{2\sqrt{3}}(2s_1 + 2s_4 - s_2 - s_5 - s_3 - s_6) \end{pmatrix} \tag{3.6.12}$$

The result is the same as in Chap. 7 of the book of Ballhausen [9] and the reader can compare two methods.

(c) T_{1u}-orbitals. IR T_{1u} is the matrix of transformation of coordinate axes and reduction coefficients can be obtained from qualitative considerations. For C_{4v} symmetry with the axis of C_4–rotations being z–axis T_{1u} matrix consists of two blocks. The upper 2×2 block corresponds to transformation of x– and y–axes, and the lower one-dimensional diagonal block of T_{1u} corresponds to transformation of z–axis and equals to one for all C_{4v} elements. Since in C_{4v} group transformations of s–function are the same as of z–function, we obtain that the column of reduction coefficients is written as $\left(0, 0, 1/\sqrt{2}\right)^T$. (Note that in this matrix z–coordinate corresponds to the last row, but in our notations s_1-orbital is in z–direction of the octahedron.) Hence, it is clear that s_1 is projecting on the third row of IR T_{1u}. Transforming this column by matrices of IR T_{1u} for rotations C_3 and C_3^2 about axis (111), we obtain contributions of s_2 and s_3

$$\overset{C_3}{\begin{pmatrix} 0 & 0 & 1 \\ 1 & 0 & 0 \\ 0 & 1 & 0 \end{pmatrix}} \begin{pmatrix} 0 \\ 0 \\ 1/\sqrt{2} \end{pmatrix} = \begin{pmatrix} 1/\sqrt{2} \\ 0 \\ 0 \end{pmatrix} \tag{3.6.13}$$

$$\overset{C_3^2}{\begin{pmatrix} 0 & 1 & 0 \\ 0 & 0 & 1 \\ 1 & 0 & 0 \end{pmatrix}} \begin{pmatrix} 0 \\ 0 \\ 1/\sqrt{2} \end{pmatrix} = \begin{pmatrix} 0 \\ 1/\sqrt{2} \\ 0 \end{pmatrix} \tag{3.6.14}$$

The remaining coefficients are obtained by inversion. Due to the oddness of the IR T_{1u}, all these coefficients are opposite in sign. Collecting all the coefficients, we get a linear combination of T_{1u} symmetry:

$$T_{1u} : \frac{1}{\sqrt{2}} \begin{pmatrix} s_3 - s_6 \\ s_2 - s_5 \\ s_1 - s_4 \end{pmatrix} \tag{3.6.15}$$

Example 3.6 Find π–orbitals, belonging to IR T_{1u} in octahedral symmetry. In this case, the orientation of axes is essential and we fix the axes as shown in Fig. 3.5a, which corresponds to our choice of left coset representatives. If the matrices of T_{1u} are taken as matrices transforming the coordinate axes of the entire molecule, the x_1 and y_1 vectors belong to its first and second rows and the column of reduction coefficients for the first center may be written in non-normalized form as $(1, 1, 0)^T$. Transforming the basis of the first center by the matrix of left coset representative we obtain the basis of the second center:

$$\overset{C_3}{\begin{pmatrix} 0\,0\,1 \\ 1\,0\,0 \\ 0\,1\,0 \end{pmatrix}} \begin{pmatrix} x_1 \\ y_1 \\ 0 \end{pmatrix} = \begin{pmatrix} 0 \\ x_2 \\ y_2 \end{pmatrix}$$

In this relation, we used the axes choice of Fig. 3.5a, according to which the action of left coset representative C_3 induces transformation $x_1 \longrightarrow x_2$ and $y_1 \longrightarrow y_2$. Similar relation for the other left coset representative is written as

$$\overset{C_3^2}{\begin{pmatrix} 0\,1\,0 \\ 0\,0\,1 \\ 1\,0\,0 \end{pmatrix}} \begin{pmatrix} x_1 \\ y_1 \\ 0 \end{pmatrix} = \begin{pmatrix} y_3 \\ 0 \\ x_3 \end{pmatrix}$$

Since the IR T_{1u} is odd, the signs of factors before coordinated connected by space inversion are equal to -1. Thus, we obtain the following basis set:

$$T_{1u} : \frac{1}{2} \begin{pmatrix} x_1 + y_3 - x_4 - y_6 \\ y_1 + x_2 - y_4 - x_5 \\ y_2 + x_3 - y_5 - x_6 \end{pmatrix} \tag{3.6.16}$$

which can be transformed to the formula of [9] by renaming of coordinate axes.

Example 3.7 For T_d point group find MO of T_2 symmetry constructed from $s-$atomic orbitals (see Fig. 3.6).

The matrices of IR T_2 for nontrivial elements of local symmetry group C_{3v} are as follows:

$$\overset{\sigma_x}{\begin{pmatrix} 1\,0\,0 \\ 0\,0\,1 \\ 0\,1\,0 \end{pmatrix}} \overset{\sigma_y}{\begin{pmatrix} 0\,0\,1 \\ 0\,1\,0 \\ 1\,0\,0 \end{pmatrix}} \overset{\sigma_z}{\begin{pmatrix} 0\,1\,0 \\ 1\,0\,0 \\ 0\,0\,1 \end{pmatrix}} \overset{C_3}{\begin{pmatrix} 0\,1\,0 \\ 0\,0\,1 \\ 1\,0\,0 \end{pmatrix}} \overset{C_3^2}{\begin{pmatrix} 0\,0\,1 \\ 1\,0\,0 \\ 0\,1\,0 \end{pmatrix}},$$

When projecting s_1-orbitals on the rows we obtain the same values for all rows. Thus the column of the coefficients of s_1 in non-normalized form is written as

$$\begin{pmatrix} 1 \\ 1 \\ 1 \end{pmatrix}$$

The element C_{2z} transforms center 1 into center 4 and the column for s_4 is obtained by multiplying the above column by the matrix $T_2(C_{2z})$:

$$\overset{C_{2z}}{\begin{pmatrix} 0 & -1 & 0 \\ -1 & 0 & 0 \\ 0 & 0 & 1 \end{pmatrix}} \begin{pmatrix} 1 \\ 1 \\ 1 \end{pmatrix} = \begin{pmatrix} -1 \\ -1 \\ 1 \end{pmatrix}$$

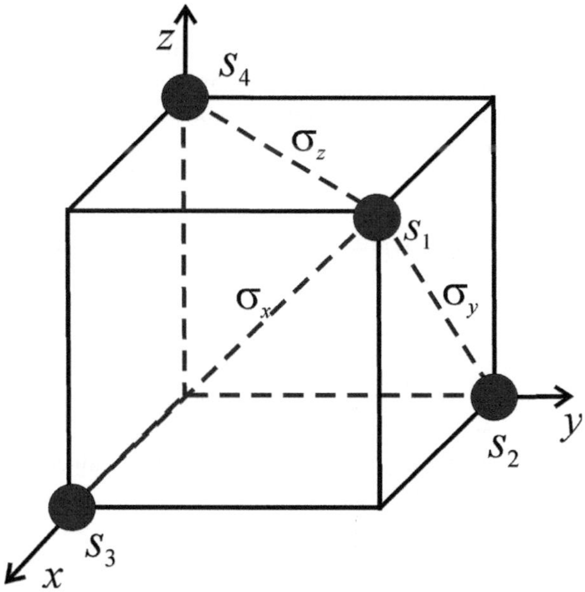

Fig. 3.6 Basis of s-orbitals in T_d symmetry

Column of the coefficients for s_2-orbital by multiplying by matrix $T_2(C_{2y})$:

$$\overset{C_{2y}}{\begin{pmatrix} 0 & 0 & -1 \\ 0 & 1 & 0 \\ -1 & 0 & 0 \end{pmatrix}} \begin{pmatrix} 1 \\ 1 \\ 1 \end{pmatrix} = \begin{pmatrix} -1 \\ 1 \\ -1 \end{pmatrix}$$

Finally, we obtain for coefficients for s_3-orbital:

$$\overset{C_{2x}}{\begin{pmatrix} 1 & 0 & 0 \\ 0 & 0 & -1 \\ 0 & -1 & 0 \end{pmatrix}} \begin{pmatrix} 1 \\ 1 \\ 1 \end{pmatrix} = \begin{pmatrix} 1 \\ -1 \\ -1 \end{pmatrix}$$

Summing and normalizing the results, we obtain

$$T_2 : \frac{1}{2} \begin{pmatrix} s_1 - s_2 + s_3 - s_4 \\ s_1 + s_2 - s_3 - s_4 \\ s_1 - s_2 - s_3 + s_4 \end{pmatrix}, \tag{3.6.17}$$

3.7 Additional Quantum Numbers for MO

It is easy to see that for a fixed total symmetry group G and a decrease of the local symmetry group H (an increase of the number of equivalent atoms in the orbit), the number of possible IRs increases and repeating IRs may appear in MO basis set. To separate the basis set in the case of repeated IRs, additional quantum numbers are required. If there is a physically distinguished intermediate group F such that $H \subset F \subset G$, then additional quantum numbers can be obtained based on the transitivity of induction theorem, according to which the representations obtained by the induction from group H directly into G and through the intermediate subgroup F are equivalent:

$$D^k \uparrow G \propto (D^k \uparrow F) \uparrow G \tag{3.7.1}$$

The induced representation on the left-hand side decomposes into the sum of the IRs Γ^q of the group G:

$$D^k \uparrow G = \sum_q f_{(k)q} \Gamma^q \tag{3.7.2}$$

Among the expansion coefficients $f_{(k)q}$, there may be numbers greater than one and our aim is to label repeating IRs by the physical additional quantum numbers. To this end, we expand the result of the first induction on the right-hand side in terms of the IRs B^i of the group F:

$$D^k \uparrow F = \sum_i f'_{(k)i} B^i \tag{3.7.3}$$

Next, we will induce separately each IR B^i in G and expand the result in terms of the IRs of the group G:

$$B^i \uparrow G = \sum_q f''_{(i)q} \Gamma^q \tag{3.7.4}$$

According to the transitivity of induction theorem, the results obtained in two ways are the same:

$$f_{(k)q} = \sum_i f'_{(k)i} f''_{(i)q} \tag{3.7.5}$$

If the intermediate expansion coefficients $f'_{(k)i}$ and $f''_{(i)q}$ do not exceed unity, then each of the repeated IRs will get an additional quantum number—the IR index of the intermediate group F. If this group is physical, then the additional quantum numbers will be a "good" quantum number, i.e. will be related to the total energy.

Example 3.8 Additional quantum numbers for the orbit of 24 atoms in O_h group.

Consider a cluster of symmetry O_h, consisting of three orbits, namely, central atom of type A, 6 atoms of type B in octahedral coordination and 24 atoms of type C, located in equivalent positions at the intersections of the coordinate planes with the faces of the cube (see Fig. 3.7). To obtain symmetrized combinations of $s-$orbitals of atoms of type C, it is necessary to induce a IR A_1 of the group C_s into the full symmetry group O_h. Making use of Frobenius reciprocity theorem, we obtain

$$A_{1(\mathsf{C}_s)} \uparrow O_h = A_{1g} + A_{2g} + 2E_g + T_{1g} + 2T_{1u} + T_{2g} + 2T_{2u}$$

In this decomposition, three IRs are present twice, and additional quantum numbers are required to label physically different basis sets of the same total symmetry. We choose C_{4v}, i.e. the local symmetry group of atoms B as an intermediate group. The induction of totally symmetric IR A_1 of C_s into C_{4v} results in

$$A_{1(C_s)} \uparrow C_{4v} = A_1, +B_1 + E \tag{3.7.6}$$

Further induction into the group O_h results in following:

$$A_{1(C_{4v})} \uparrow O_h = A_{1g} + E_g + T_{1u} \tag{3.7.7}$$

$$B_{1(C_{4v})} \uparrow O_h = A_{2u} + E_g + T_{2u} \tag{3.7.8}$$

$$E_{(C_{4v})} \uparrow O_h = T_{1g} + T_{1u} + T_{2g} + T_{2u} \tag{3.7.9}$$

Thus, by the application of the induction through the intermediate subgroup repeated IRs acquire additional quantum numbers—indices of IRs of the intermediate subgroup and may be written as $E_g(A_1)$, $E_g(B_1)$, $T_{1u}(A_1)$, $T_{1u}(E)$, $T_{2u}(B_1)$ and $T_{2u}(E)$. In order to envisage the physical meaning of these additional quantum numbers, we construct some such symmetrical combinations. The numbering of atoms (see Fig. 3.7) is as follows. The numbers 1, 2, 3, 4, 5 and 6 correspond to the vertices $(z,\ x,\ y,\ -z,\ -x,\ -y)$ of the octahedron. The atoms on the top face of the cube are labeled α_1, β_1, γ_1 and δ_1. We choose as the left coset representatives in the decomposition of O_h with respect to C_{4v} the rotations C_3 and C_3^2 about the (111) axis, the spatial inversion, and the products of these rotations and inversion. These elements of the group transform the functions $\left(s_{\alpha 1},\ s_{\beta 1}, s_{\gamma 1}, s_{\delta 1}\right)$ into the corresponding functions $\left(s_{\alpha i},\ s_{\beta i}, s_{\gamma i}, s_{\delta i}\right)$ on the other faces of the cube. These functions are basis sets on the local groups isomorphic to C_{4v}.

As an example, we will construct bases of symmetry E_g of the group O_h from functions that have symmetry A_1 and B_1 on the group to C_{4v}. The symmetrized on C_{4v} group combinations of $s-$functions on the upper face are written as

$$\varphi_1(A_1) = \frac{1}{2}\left(s_{\alpha 1} + s_{\beta 1} + s_{\gamma 1} + s_{\delta 1}\right) \tag{3.7.10}$$

$$\varphi_1(B_1) = \frac{1}{2}\left(s_{\alpha 1} - s_{\beta 1} + s_{\gamma 1} - s_{\delta 1}\right) \tag{3.7.11}$$

It was shown in a previous section that when projecting basis set belonging to IR A_1 of C_{4v} on IR E_g of O_h we obtain the column $(0, 1/\sqrt{3})^T$ (3.6.9). Since the $s-$basis sets on the other centers are obtained by the left coset representatives, symmetrized combinations $\varphi_i(A_1)$ and $\varphi_i(B_1)$ are obtained from formulas (3.7.10) and (3.7.11) by replacing 1 with i. Obviously, symmetrized combinations E_g with

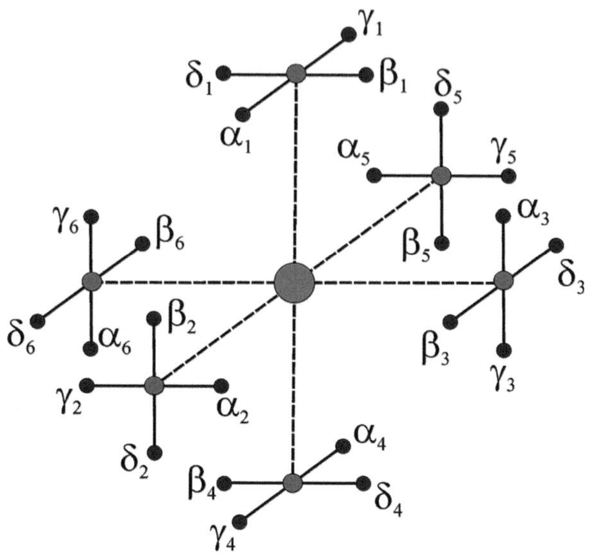

Fig. 3.7 24-fold coordinations of atoms in O_h symmetry

an additional quantum number A_1 will have same coefficients before $\varphi_i(A_1)$ as that before six s_i-orbitals in O_h symmetry (3.6.12) (see Table 3.6).

Basis set of E_g symmetry with additional quantum number B_1 is easily obtained making use of projection technique described in the previous section. Projection of $\varphi_1(B_1)$ on the rows of E_g results in column $(1/\sqrt{3}, 0)^T$. To find linear combinations for other centers, it is necessary to transform the obtained column by the matrices E_g of the elements which transform the basis of the first center into the basis of the second and third centers:

$$\overset{C_3}{\begin{pmatrix} -\frac{1}{2} & \frac{\sqrt{3}}{2} \\ \frac{-\sqrt{3}}{2} & -\frac{1}{2} \end{pmatrix}} \begin{pmatrix} 1/\sqrt{3} \\ 0 \end{pmatrix} = \begin{pmatrix} -\frac{1}{2\sqrt{3}} \\ -\frac{1}{2} \end{pmatrix}$$

$$\overset{C_3^2}{\begin{pmatrix} -\frac{1}{2} & \frac{-\sqrt{3}}{2} \\ \frac{\sqrt{3}}{2} & -\frac{1}{2} \end{pmatrix}} \begin{pmatrix} 1/\sqrt{3} \\ 0 \end{pmatrix} = \begin{pmatrix} -\frac{1}{2\sqrt{3}} \\ \frac{1}{2} \end{pmatrix}$$

Due to the parity of the IR E_g, the coefficients for the bases related by inversion are the same. Summing up all contributions, we obtain linear combinations for the IR E_g, which differ by additional quantum numbers A_1 and B_1 (i.e. labels of IRs of an intermediate group C_{4v}), presented in Table 3.6.

The case which occurs depends on the type of orbitals located at the vertices B_i of the octahedron. If there are $s-$orbitals there will be case $E_g(A_1)$ and if there are $d-$orbitals there will be case $E_g(B_1)$.

Table 3.6 Molecular orbitals E_g with additional quantum numbers A_1 and B_1 constructed on the basis of $s-$orbitals in 24-fold coordination in O_h symmetry (see Fig. 3.7). Functions $\varphi_i(A_1)$ and $\varphi_i(B_1)$ are defined by formulas (3.7.10) and (3.7.11)

$E_g(A_1)$	$\frac{1}{2}[\varphi_2(A_1)-\varphi_3(A_1)+\varphi_5(A_1)-\varphi_6(A_1)]$
	$\frac{1}{2\sqrt{3}}[2\varphi_1(A_1)+2\varphi_4(A_1)-\varphi_2(A_1)-\varphi_5(A_1)-\varphi_3(A_1)-\varphi_6(A_1)]$
$E_g(B_1)$	$\frac{1}{2\sqrt{3}}[2\varphi_1(B_1)+2\varphi_4(B_1)-\varphi_2(B_1)-\varphi_5(B_1)-\varphi_3(B_1)-\varphi_6(B_1)]$
	$\frac{1}{2}[-\varphi_2(B_1)+\varphi_2(B_1)-\varphi_4(B_1)+\varphi_5(B_1)]$

3.8 Irreducibility of Induced Representations

The reducibility or irreducibility of the induced representation $D_{(H)} \uparrow G$ is closely related to the structure of the group with respect to the subgroup H and to the properties of the IR D. Of course, in each particular case, the induced representation can be easily decomposed into IRs using the Frobenius reciprocity theorem. However, since the irreducibility property of the representation induced from the wavevector group in crystals is of great importance in solid-state physics, these properties will be briefly considered in the present section. Let H be a subgroup of G. A normalizer of a subgroup H is a subgroup $N \subset G$ such that the group transformed from H by actions of elements n such that $n \subset N$ coincides with H:

$$^{n}H = nHn^{-1} = H \tag{3.8.1}$$

It is obvious that the inverse element n^{-1} has the same property, the conjugation of which is denoted by an overbar:

$$^{\bar{n}}H = n^{-1}Hn = H \tag{3.8.2}$$

All elements of H and the inversion, which commutes with all group elements, belong to the normalizer N. If H is an invariant subgroup, then all elements of G have the properties of normalizer's elements. Since the conjugate subgroup ^{n}H is the same as H, for any IR D of the subgroup H, one can define the conjugate representation $_{n}D$ as follows:

$$_{n}D(h) = D(^{\bar{n}}h) = D(n^{-1}hn) \tag{3.8.3}$$

Note that this definition of $_{n}D$ coincides with the definition of the representation of the conjugate subgroup, which makes sense even if the conjugate subgroup $^{n}H = nHn^{-1}$ does not coincide with the original one. The conjugate representations $_{n}D$ are generally not equivalent, but in some cases they turn out to be equivalent. It can be shown that all elements k of the normalizer N for which the equivalence relation holds

$$_{k}D(h) =_{\bar{k}} D(h) \sim D(h) \tag{3.8.4}$$

form a group K called the little group of IR D, which will be denoted by $K(D)$.

A conjugation by an element $h \in H$ transforms the element h_i into an element h_j, which belongs to the same class of the group H. The characters of elements of the same class are the same, so all elements of H belong to the little group $K(D)$. The set of all non-equivalent representations transformed from a representation D by elements of normalizer N forms a superorbit $S(D)$:

$$S(D) = \sum_{\nu} {}_{\nu}D = \{{}_{\nu}D\} \quad (3.8.5)$$

Each of the non-equivalent IRs ${}_{\nu}D$ is called a prong of this superorbit. The definition of a little group can be applied to each of the prongs of the superorbit. It can be shown that the little group of the prong ${}_{\nu}D$ of the superorbit $S(D)$ is obtained by conjugating the little group of the original ray $K(D)$ by the element ν:

$$K({}^{\nu}D) = [K(D)]^{\nu} = \nu K(D)\nu^{-1} \quad (3.8.6)$$

The definition of a superorbit has the disadvantage that each of the prongs is taken into account several times. Let us decompose the normalizer into left cosets of the little group:

$$N = \sum_{\lambda} s_{\lambda} K(D) \quad (3.8.7)$$

Obviously, conjugation by all elements of the left coset defined by one element s_{λ} will give the same prong ${}_{\lambda}D$. Therefore, to obtain all the rays that make up the orbit of the representation D, it suffices to have representatives of the left cosets in the expansion of N in K. In what follows, we assume that H is an invariant subgroup. In this case, the whole group G is a normalizer.

Let us consider two theorems related to the construction of the IRs for space groups of crystals. The complete proof of this theorems is presented in a book of Altmann [20].

Irreducibility Theorem. Let T be an invariant subgroup of G, d^{τ} be an irreducible representation of T, and K be a little group of IR d^{τ}. (Note that $T \subset K^{\tau} \subset G$.) Let the IR $F^{\tau,i}$ of a little group K be known. Then the induced representation

$$F^{\tau i} \uparrow G \quad (3.8.8)$$

will be irreducible on the group G. Here i denotes IRs of K.

Completeness Theorem. The set of representations $F^{\tau,i} \uparrow G$ where τ runs over all IRs of T and i runs through all IRs of K^{τ} forms a complete set of IRs of the group G.

The mathematical theory discussed in this section is directly related to the space groups of crystals. Here G is the space group, T is an invariant subgroup of translations that has one-dimensional representations, determined by the wavevector k, a little group K is a subgroup of the space group that does not change the wavevector.

It follows from the irreducibility theorem that IRs of the wavevector group induced into the space group are irreducible representation. It follows from the completeness theorem that in order to get all IRs of the space group, one should consider all non-equivalent IRs of translation group, i.e. when k runs over all Brillouin zone.

Example 3.9 Irreducibility of induced representations of group T.

The group T (group of rotations of tetrahedron) has an invariant subgroup D_2. Therefore, the whole group T is a normalizer of D_2. It is easy to verify that when conjugated by the elements of the group D_2 itself, equivalent IRs are obtained. The left coset representatives in the decomposition of T with respect to D_2 are the elements $\{E, C_3, C_3^2\}$. The representations of group D_2 and those conjugated by the left coset representatives are presented in Table 3.7. The elements $n^{-1}h_i n$ conjugated to h_i (second line) by the elements $n = E$, C_3 and C_3^2 are presented in the third line of Table 3.7. Since the conjugation is performed by the formula:

$$_nD(h) = D(n^{-1}hn) \tag{3.8.9}$$

we associate the element h_i in conjugate representation with the character of the original representation for the element $n^{-1}h_i n$. Characters of all representations of D_2, conjugated by the three coset representatives, are presented in last three lines. Note that these characters correspond to the elements listed in the second line.

It is seen from Table 3.7 that the representations $_nA$ conjugated to A are equivalent to the original one. Therefore, the small group for A is the entire group T and the induction of A from T into T (i.e. identity operation) results in irreducible representation A of T. On the other hand, as it can be immediately verified making use of characters of T_h presented in Table 38 of Chap. 4, the induction of A from D_2 into T results in a reducible representation of T:

$A_{(D_2)} \uparrow G = 3A$

It is seen from Table 3.7 that representations conjugated to B_1, B_2 and B_3 are not equivalent to original ones. For example, the representations conjugated to B_1 by C_3 and C_3^2 are IRs B_3 and B_2, respectively. Hence it follows that the little group for IRs

Table 3.7 Data for the investigation of induced representation of group T. The elements of group D_2 are listed three times in the second line. The elements conjugated by n (in the first row) are placed in the third row under the original elements. Conjugated representations are in the last four rows

D_2	$n = E$				$n = C_3^2$				$n = C_3$			
h	E	C_{2x}	C_{2y}	C_{2z}	E	C_{2x}	C_{2y}	C_{2z}	E	h_2	C_{2y}	C_{2z}
$n^{-1}h_i n$	E	C_{2x}	C_{2y}	C_{2z}	E	C_{2y}	C_{2z}	C_{2x}	E	C_{2z}	C_{2x}	C_{2y}
$_nA(h)$	1	1	1	1	1	1	1	1	1	1	1	1
$_nB_1(h)$	1	−1	−1	1	1	−1	1	−1	1	1	−1	−1
$_nB_2(h)$	1	−1	1	−1	1	1	−1	−1	1	−1	−1	1
$_nB_3(h)$	1	1	−1	−1	1	−1	−1	1	1	−1	1	−1

B_i is the group D_2 and induced representations $B_{i(D_2)} \uparrow G$ are irreducible. It can be easily obtained making use of Frobenius reciprocity theorem that

$B_{i(D_2)} \uparrow G = T,\ i = 1, 2, 3.$

The subduction of T on D_2 results in three non-equivalent representations:

$T \downarrow D_2 = B_1 + B_2 + B_3$

Thus we obtained that the induction of IRs of small group (A of T and B_1 of D_2) into T results in irreducible representations A and T, respectively, of the whole group T.

3.9 Space Groups

In this and in the next sections, we briefly consider the theory of space groups based on the induced representations. The complete theory of space group IRs and tables of space group are presented in books of Bradley and Cracknell [19] and Melnikov and Reser [61] and Kovalev [60]. The applications of induced representation theory make possible to obtain compact formulas for the calculation of selection rules, compatibility relations and the structure of two-electron states in crystal solids. These applications will be considered in subsequent sections and in Chap. 4.

The space groups G include operations of three types: translations onto the lattice vector $\{E|t\}$, $t \in T$, operations of point groups (rotations, reflections and inversion) $\{s|0\}$, and screw axes and glide plane elements $\{\nu|\tau_\nu\}$, $\tau_\nu \notin T$. In these elements called non-symmorphic, each point group ν element appears with its improper translation τ_ν. The space-group element in Seitz notation is written as $\{\alpha|\tau_\alpha + t\}$, where symbol before vertical bar corresponds to point group element and the symbols after vertical bar correspond to translation.

The multiplication of space-group elements is written as

$$\{\alpha|\nu\}\{\beta|\mu\} = \{\alpha\beta|\nu + \alpha\mu\} \tag{3.9.1}$$

The inverse element is defined as

$$\{\alpha|\nu\}^{-1} = \left\{\alpha^{-1}| - \alpha^{-1}\nu\right\} \tag{3.9.2}$$

The identity element is written as $\{E|0\}$. Point group operations form a subgroup P:

$$\{s|0\}\{p|0\} = \{sp|0\},\ s, p \in P \tag{3.9.3}$$

Translations also form a subgroup T which is invariant subgroup of G:

$$\{E|t_1\}\{E|t_2\} = \{E|t_1 + t_2\},\ t_1, t_2, t_1 + t_2 \in T \tag{3.9.4}$$

However, rotations with improper translations $\{\nu|\tau_\nu\}$ do not form a set that is closed under multiplication. The product of two such elements is given by the formula:

$$\{\nu|\tau_\nu\}\{\mu|\tau_\mu\} = \{\nu\mu|\tau_\nu + \nu\tau_\mu\} \quad (3.9.5)$$

As a result, there can be both an element of the subgroup of pure rotations with translation onto the lattice vector, and a rotation with improper translation. Also the elements $\{\nu|\tau_\nu\}$ are called improper rotations. The translation group defines the Bravais lattice which is invariant under point group operations.

It is easy to see that elements with improper translations do not commute with elements of the translation subgroup. However, the translation group T is an invariant subgroup of G:

$$\{q|\tau_q\}\{E|t\}\{q|\tau_q\}^{-1} = \{q|\tau_q + qt\}\{q^{-1}| - q^{-1}\tau_q\} = \{E|\tau_q + qt - \tau_q\} = \{E|qt\} \quad (3.9.6)$$

The space group can be represented as a decomposition into left cosets in terms of the invariant subgroup of translations:

$$G = \sum_p \{p|\tau_p\}\, T \quad (3.9.7)$$

In this decomposition, each element of the point group corresponds to a defined improper translation or zero translation. The representatives of the left cosets themselves do not form a group. The group is formed by a set of pure rotations, supplemented by the point group parts of the screw axes and glide planes. These elements, called the central extension, form a group denoted by $\hat{G}$. Note that $\hat{G}$ is not a subgroup of G. If the space group is symmorphic, i.e. does not contain screw axes and glide planes, then the decomposition into the left cosets is written as

$$G = \sum_p \{p|0\}\, T \quad (3.9.8)$$

In this case, elements $\{p|0\}$ form a group. The general element $\vec{t}$ of a translation group is written in terms of translations along the lattice basis vectors a, b and c:

$$\vec{t} = n_1 a + n_2 b + n_3 c \quad (3.9.9)$$

The translation group is equal to the direct product of three subgroups of one-dimensional translations:

$T = T_a \otimes T_b \otimes T_c$

All elements of the one-dimensional group T_a are expressed as powers of the primitive translation $\{0|t_a\}^n$. The rule for multiplying elements of a translation group has the form:

$$\{0|t_a\}^n\{0|t_a\}^m = \{0|t_a\}^{m+n} \quad (3.9.10)$$

The translation group T_a is formally infinite, but can be reduced to a finite one by introducing periodic boundary conditions for each of its one-dimensional subgroups:

$$T_a = \{E|t_a\} + \{E|t_a\}^2 +, \{E|t_a\}^{n_1} = \{E|0\} \tag{3.9.11}$$

Taking into account similar relations for two other one-dimensional translations, we get that the number of elements in the three-dimensional group of translations is equal to

$$|T| = n_1 n_2 n_3 \tag{3.9.12}$$

Then the number of elements in the entire space group is as follows:

$$|G| = |T| \left|\hat{G}\right| \tag{3.9.13}$$

The translation group is commutative (Abelian) and each of its elements forms a separate class, so the number of classes of the group is equal to the number of elementary cells of the crystal N. The IR of a one-dimensional translation group is written as

$$d^{k_a}\left(\{0|t_a\}\right) = \exp(-ik_a t_a) \tag{3.9.14}$$

For a three-dimensional group of translations, reciprocal lattice vectors are defined as follows:

$$k_a = \frac{2\pi\,[b \times c]}{a \cdot [b \times c]}. \tag{3.9.15}$$

Similar relations for k_b and k_c are obtained cyclic permutations. Scalar products of k_a by the lattice vectors are written as

$$k_a \cdot a = 2\pi \tag{3.9.16}$$

$$k_a \cdot b = 0 \tag{3.9.17}$$

$$k_a \cdot c = 0 \tag{3.9.18}$$

The formulas for the remaining reciprocal lattice vectors are obtained by cyclic permutations.

Formula (3.9.14) for the IR of one-dimensional translation group may be extended to a three-dimensional case by introducing a three-dimensional wavevector:

$$\vec{k} = \alpha k_a + \beta k_b + \gamma k_b \tag{3.9.19}$$

where α, β and γ are real numbers. Also we define a reciprocal lattice vector as

$$\vec{g} = lk_a + mk_b + nk_b \quad (3.9.20)$$

where l, m and n are integer numbers.

If two vectors of reciprocal space differ by a reciprocal lattice vector:

$$\vec{k} = \vec{k}' + \vec{g} \quad (3.9.21)$$

then the product $\vec{g}\vec{t}$ of a reciprocal lattice vector by any translation group element equals to the integer number of 2π and the corresponding IRs $d^{\vec{k}}$ and $d^{\vec{k}'}$ of translation group coincide. It should be noted that the quantity $\hbar\vec{k}$ in a solid is an electron momentum, and wavevectors $\vec{k}$ and $\vec{k}'$ equivalent, but not equal to each other. Therefore, in order to obtain all IRs, it suffices to consider only a segment of length $|k_i|$ in each direction in the reciprocal lattice. Usually a segment between two equivalent vectors $-k_i/2$ and $k_i/2$ is taken and planes perpendicular to the vectors are drawn through their ends. Having done this for all three vectors of the reciprocal lattice, we obtain a polyhedron in reciprocal space, called the Brillouin zone (BZ), or the first Brillouin zone. By performing shifts on the vectors of the reciprocal lattice, one can completely fill the reciprocal space. The IR of the translation group is determined by the points inside the first Brillouin zone. However, to construct the IRs of the space group, it is necessary to consider the wavevector symmetry group H and its IRs, what will be done in the next section.

3.10 Irreducible Representations of Space Groups

As shown in the previous section, the representation $d^{\vec{k}}$ of a translation group is characterized by the wavevector $\vec{k}$ and is given by

$$d^{\vec{k}}\left\{E|\vec{t}\right\} = \exp(-i\vec{k}\cdot\vec{t}) \quad (3.10.1)$$

In what follows, we will omit the vector notations for the Bravais and reciprocal lattices.

For each IR d^k of an invariant translation subgroup T, one can define a representation ${}_gd^k$ conjugated by an element g of the entire group G:

$${}_gd^k(t) = d^k(g^{-1}tg) \quad (3.10.2)$$

For the result of conjugation of the IR d^k of the translation group T by an arbitrary element $\{\alpha|\tau\}$ of the space group, we can write

$${}_{\{\alpha|\tau\}}d^k\{E|t\} = d^k\{\alpha|\tau\}^{-1}\{E|t\}\{\alpha|\tau\} =$$

$$d^k \{\alpha^{-1}| - \alpha^{-1}\tau\} \{\alpha|\tau + t\} = d^k \{E| - \alpha^{-1}\tau + \alpha^{-1}\tau + \alpha^{-1}t\} =$$

$$\exp\left(-ik \cdot \alpha^{-1}t\right) = \exp\left(-i\alpha k \cdot t\right) = d^{\alpha k} \{E|t\} \tag{3.10.3}$$

If $\{\psi_k\}$ is the basis of the original IR, then the basis of the conjugate IR will be the functions $\{\psi_{\alpha k}\}$. Using all elements of the group G for the conjugation operation, we obtain the superorbit of the representation d^k and the corresponding superstar of the vector k, which consists of the vectors αk, called prongs. However, in a superstar, prongs occur more than once. Our task is to find a star of vector k, denoted by $\{k\}$, in which each prong appears only once. Obviously, since translations on the lattice vector do not affect the conjugation, instead of the entire group G, it suffices to take the elements of the point group elements p, such that $p \in \hat{G}$. If the wavevector is in the symmetrical direction in the Brillouin zone, then rotations about this direction and reflections in planes passing through this axis do not change the wavevector. In addition, on the boundary of the Brillouin zone, cases are possible when the rotation transforms k into an equivalent vector that differs by the reciprocal lattice vector. Thus we arrive at the following definition. The little group H of the wavevector k is a group consisting of elements $\{p|\tau_p + t\}$, whose rotations p transform k into an equivalent vector:

$$p\vec{k} = \vec{k} + \vec{g} \tag{3.10.4}$$

where $\vec{g}$ is a reciprocal lattice vector.

In group theory, it is proved that if for an arbitrary invariant group T, an irreducible representation D^k of its little group H is obtained, then induction of the IR D^k of the little group results in an irreducible representation of the entire group. It follows that if the IR D^k of the group H of the wavevector k is known, then the induced representation $D^k \uparrow G$ will be the IR of the space group. Thus, the IR of the space group is determined by the wavevector k, the group H of the wavevector (little group) and the IR D^k of the little group (small IR).Thus, the IR of the space group is the induced representation defined by the formula:

$$D^k \uparrow G(g) = \begin{cases} D^k(s_i^{-1} g s_j), \text{ if } s_i^{-1} g s_j \in H \\ 0, \text{ if } s_i^{-1} g s_j \notin H \end{cases} \tag{3.10.5}$$

where are s_i are left coset representatives in the decomposition of G: $G = \sum_i s_i H$.

When writing translation elements explicitly $s_i = \{\hat{s}_i|\tau_i\}$ and $g = \{\hat{g}|\tau + t\}$, the element of the block of the induced representation matrix may be written as

$$D^k(s_i^{-1} g s_j) = D^k\left(\left\{\hat{s}_i^{-1}| - \hat{s}_i^{-1}\tau_i\right\}\{\hat{g}|\tau + t\}\{\hat{s}_j|\tau_j\}\right) =$$
$$D^k\left(\left\{\hat{s}_i^{-1}\hat{g}| - \hat{s}_i^{-1}\tau_i + \hat{s}_i^{-1}(\tau + t)\right\}\{\hat{s}_j|\tau_j\}\right) =$$
$$D^k\left(\left\{\hat{s}_i^{-1}\hat{g}\hat{s}_j| - \hat{s}_i^{-1}\tau_i + \hat{s}_i^{-1}(\tau + t) + \hat{s}_i^{-1}\hat{g}\tau_j\right\}\right) = D^{\hat{s}_i k}\left(\left\{\hat{s}_i^{-1} g\hat{s}_j| - \tau_i + (\tau + t) + g\tau_j\right\}\right) \tag{3.10.6}$$

Hence, it follows that prongs of the wavevector star correspond to the rows of the induced representation.

3.11 Projective Representations of the Wavevector Group

The little group H has an infinite order (reducible to finite, but very large by introducing periodic boundary conditions). For simplification, a little factor group H/T is introduced.

$$H/T = \sum_p \{p|\tau_p\} \tag{3.11.1}$$

This group consists of rotational elements $\{\alpha|0\}$ of the wavevector group, and some elements $\{\alpha|\tau_\alpha\}$ (which we will call non-symmorphic) enter with their own improper translations, which are considered fixed. Such a set of elements is not closed under multiplication (except for the case of symmorphic groups). The operation of multiplication in a factor group is understood as the multiplication of the left cosets:

$$[\{\alpha|\tau_\alpha\}T]\left[\{\beta|\tau_\beta\}T\right] = \left[\{\gamma|\tau_\gamma\}T\right],\ \alpha\beta = \gamma \tag{3.11.2}$$

In order not to work with left cosets, a small co-group $\bar{H}$ is introduced, consisting of the rotational parts of p elements $\{p|\tau_p + t\}$ belonging to the group H of the wavevector k.

Let us decompose the space group into the left cosets in terms of the wavevector group:

$$G = \sum_i \{s_i|\tau_{s_i}\} H \tag{3.11.3}$$

It is obvious that the rotational parts of the representatives of the left cosets are the representatives of the left cosets in the decomposition of the point group $\bar{G}$ corresponding to the space group G with respect to the little co-group:

$$\bar{G} = \sum_i s_i \bar{H} \tag{3.11.4}$$

The action of these representatives of the left cosets on the vector k results in a collection of vectors, called its star:

$$\{k\} = \sum_i s_i k,\ i = 1, .. \left|\bar{G}\right| / \left|\bar{H}\right| \tag{3.11.5}$$

Let $D^k(\{\alpha|\tau_\alpha\})$ be the wavevector group, then the matrices:

$$\hat{D}^k(\alpha) = \exp(ik \cdot (\tau_\alpha + t))\ D^k\{\alpha|\tau_\alpha + t\} \tag{3.11.6}$$

form a projective representation of a little co-group. It is easy to find the multiplication rule for elements of a little co-group:

$$\{\alpha|\tau_\alpha\}\left\{\beta|\tau_\beta\right\} = \left\{\gamma|\tau_\gamma + t\right\} \tag{3.11.7}$$

$$t = \tau_\alpha + \alpha\tau_\beta - \tau_\gamma \tag{3.11.8}$$

Then the multiplication of the projective representations introduced above may be obtained as follows:

$$\hat{D}^k(\alpha)\hat{D}^k(\beta) = \exp(ik \cdot \tau_\alpha)\ D^k\{\alpha|\tau_\alpha\}\exp(ik \cdot \tau_\beta)\ D^k\left\{\beta|\tau_\beta\right\} =$$

$$= \exp(ik \cdot (\tau_\alpha + \tau_\beta))\ D^k\left\{\gamma|\tau_\alpha + \alpha\tau_\beta\right\} = \exp(ik \cdot (\tau_\alpha + \tau_\beta))\ \hat{D}^k(\gamma)\exp(ik \cdot (-\tau_\alpha - \alpha\tau_\beta)) =$$

$$= \hat{D}^k(\gamma)\exp(ik \cdot (\tau_\beta - \alpha\tau_\beta)) \tag{3.11.9}$$

The factor in the last line is an element of the factor system of the projective representation. It can be transformed to a more convenient form:

$$\exp(ik \cdot (\tau_\beta - \alpha\tau_\beta)) = \exp(ik \cdot \tau_\beta - ik \cdot \alpha\tau_\beta)) = \exp(i\left(k - \alpha^{-1}k\right) \cdot \tau_\beta) \tag{3.11.10}$$

Thus the factor system of projective representation is defined as follows:

$$\omega(\alpha, \beta) = \exp(ib \cdot \tau_\beta) \tag{3.11.11a}$$

where the reciprocal lattice vector b is defined from the relation:

$$k - \alpha^{-1}k = b \tag{3.11.12}$$

It can be seen from (3.11.12) that the first factor in formula (3.11.11a) is different from unity, if α^{-1} transforms the wavevector k into a vector that differs from it by a reciprocal lattice vector b, and if improper translation of the second multiplier $\left\{\beta|\tau_\beta\right\}$

is nonzero. Let us consider the IR of the space group in the case when $g = \{E|t\}$. In this case, only the blocks on the main diagonal are nonzero, for which we have

$$\left\{p_j^{-1}|0\right\}\{E|t\}\{p_j|0\} \tag{3.11.13}$$

$$\begin{aligned} D^k \uparrow G(\{E|t\})_{j,j} &= D^k(\left\{p_j^{-1}|0\right\}\{E|t\}\{p_j|0\}) = D^k(\left\{E|\, p_j^{-1}t\right\} \\ &= \hat{D}^k(E)\exp(-ik\cdot p_j^{-1}t) = \hat{D}^k(E)\exp(-ip_jk\cdot t) \end{aligned} \tag{3.11.14}$$

The matrices $\hat{D}^k(E)$ are unit matrices, and from the obtained formula it follows that on the main diagonal of the matrix there are unit matrices multiplied by $\exp(-ik_j \cdot t)$, where k_j are the rays of the vector star $\{k\}$.

Bloch's theorem. Basis of functions $\psi_k(r)$ transforming according to the IR d^k of the translation group, i.e.:

$$\{E|t\}\,\psi_k(r) = \exp(-ik\cdot t)\psi_k(r) \tag{3.11.15}$$

may be written as

$$\psi_k(r) = \exp(ik\cdot r)u(r) \tag{3.11.16}$$

where $u(r)$ is periodic function with period equal to the lattice translation: $u(r+t) = u(r)$

The proof is as follows:

$$\{E|t\}\,\psi_k(r) = \exp(ik\cdot\{E|t\}^{-1}r)u(\{E|t\}^{-1}r) = \exp(ik\cdot(r-t))u(r) = \exp(-ik\cdot r)\psi_k(r) \tag{3.11.17}$$

Example 3.10 Build IRs for the elements $\{E|t\}$, $\{C_{2b}|t\}$ and $\{C_{4z}|t\}$ of the space group with a simple orthorhombic lattice (point group D_{4h}, see Fig. 3.8) for the wavevector k in the direction Σ. It is seen in Fig. 3.8 that in the $\Sigma-$direction the wavevector group is a group C_{2v}, whose IRs are presented in Table 3.8.

The IRs of a space group are obtained by the induction from C_{2v}. The group D_{4h} decomposes into four left cosets with respect to its subgroup C_{2v}. Representatives s_i of these left cosets are $\{E, C_{4z}, I, C_{4z}^3\}$. The action of these coset representatives on the wavevector k_1 results in its star $\{k_1, k_2, k_3, k_4\}$, as shown in Fig. 3.8.

According to formula (3.11.14), the IR for the identity rotational element and translations $t \in T$ is written as:

$$\Gamma\{E|t\} = \begin{pmatrix} e^{-ik_1t} & & & \\ & e^{-ik_2t} & & \\ & & e^{-ik_3t} & \\ & & & e^{-ik_4t} \end{pmatrix}$$

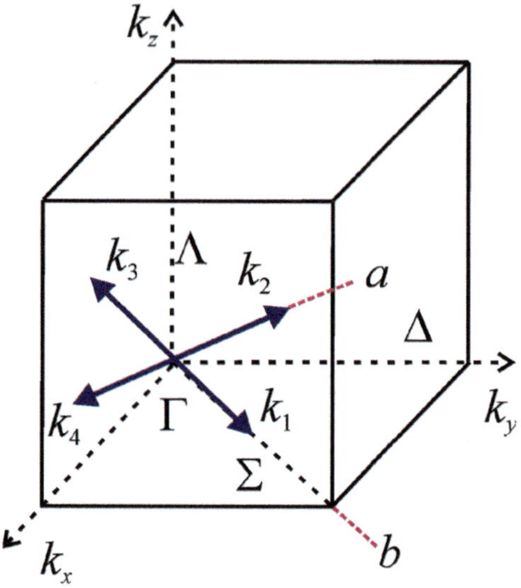

Fig. 3.8 Part of BZ for the simple Bravais lattice with point group D_{4h} and the wavevector star in Σ-direction

Table 3.8 Characters of group C_{2v}. Notations of the axes are shown in Fig. 3.8

	E	C_{2b}	σ_z	σ_a
A_1	1	1	1	1
A_2	1	1	-1	-1
B_2	1	-1	1	-1
B_1	1	-1	-1	1

To obtain nonzero blocks of matrix for C_{2b}, we need to check relations s_i^{-1} $C_{2b}s_j \in C_{2v}$ for all left coset representatives s_i and s_j. Making use of the multiplication table of the group D_{4h} we obtain following relations:

$(1, 1) : B_1(C_{2b}) = -1$
$(2, 4) : B_1(C_{4z}^3 C_{2b} C_{4z}^3) = B_1(C_{2b}) = -1$
$(3, 3) : B_1(I, C_{2b}, I) = B_1(C_{2b}) = -1$
$(4, 2) : B_1(C_{4z} C_{2b} C_{4z}) = B_1(C_{2b}) = -1$

where the numbers of rows and of columns are indicated in parenthesis.

Taking into account (3.10.5) the IR for the element $\{C_{2b}|t\}$ is written as

$$\Gamma\left\{C_{2b}|t\right\} = \begin{pmatrix} -e^{-ik_1t} & & & \\ & & & -e^{-ik_2t} \\ & & -e^{-ik_3t} & \\ & -e^{-ik_4t} & & \end{pmatrix}$$

For the element C_{4z}, the nonzero blocks are as follows:

$(1, 4) : B_1(EC_{4z}C_{4z}^3) = B_1(E) = 1$
$(2, 1) : B_1\left(C_{4z}^3 C_{4z} E\right) = B_1(E) = 1$

$(3, 2) : B_1(IC_{4z}C_{4z}) = B_1(\sigma_z) = -1$
$(4, 3) : B_1(C_{4z}C_{4z}I) = B_1(\sigma_z) = -1$
Thus the matrix for $\{C_{4z}|t\}$ is written as

$$\Gamma\{C_{4z}|t\} = \begin{pmatrix} & & & e^{-ik_1t} \\ e^{-ik_2t} & & & \\ & -e^{-ik_3t} & & \\ & & -e^{-ik_4t} & \end{pmatrix}$$

3.12 Compatibility Relations

The energy of one-electron state in a crystal depends on two quantum numbers: the wavevector k in a basis domain of a BZ, which uniquely defines the little group H, and the label κ of small IR D^{κ} of H. When the wavevector takes a path along a trajectory in a $k-$space, the little group may change, but the energy spectrum should by continuous. The compatibility relation for continuous transitions between symmetrical lines and points of a BZ was obtained in foundational work of Bouckaert, Smoluchowski and Wigner [62]. In this section, we consider the compatibility relations in the framework of induced representation method and the difference of compatibility relations between symmorphic O_h^5 and non-symmorphic O_h^7 space groups having face-centered cubic lattice is demonstrated. The BZ for a face-centered cubic lattice with standard notations of points and of symmetry is shown Fig. 3.9. Let us find the compatibility relations for the IRs when k is passing from the center of the BZ (point Γ) to the axes of the third and fourth orders (directions Λ and Δ in Fig. 3.9).

Let the wavevector be in the direction Λ and its length approaches to zero. In this case, the exponential factors in the formula for the space group IRs (3.10.6) tend to zero. In this case, the general formula for the IRs of the space group (3.10.5) becomes the formula for the induced representation of the point group $D^{\kappa}_{(H)} \uparrow O_h$. To expand it in terms of IRs Γ^q in Γ point, one can use the Frobenius reciprocity theorem, by which we get that the IR at the point Γ can pass into the IR $D^{\kappa}_{(H)} \uparrow O_h$ in some direction if the character Γ^q is not orthogonal to the character of D^{κ} of the wavevector group. The compatibility relations of IRs at point Γ with IRs in directions Λ (wavevector group C_{3v}) and Δ (wavevector group C_{4v}) are presented in Table 3.9. Note that the degeneracy of states in less symmetrical points in a BZ is larger than degeneracy in more symmetrical points. Consider the direction Δ, where the star $\{k\}$ consists of six prongs (see Fig. 3.10). Thus, the dimension of induced representation $E_{(\mathsf{C}_{4v})} \uparrow O_h$ equals to 12. It is seen from Table 3.9 that E (in $\Delta-$direction) is compatible with T_{1g}, T_{2g}, T_{1u} and T_{2u} at Γ point. Likewise, the dimension of $A_{1(\mathsf{C}_{4v})} \uparrow O_h$ equals to 6 and it is compatible with A_{1g}, E_g and T_{1u} at Γ point. In a similar way, i.e. by checking the orthogonality of characters of groups O_h and C_{3v}, relations can be obtained for the transition between the $\Gamma-$point and the $\Lambda-$direction. In this case, star consists of 8 prongs the dimensions of $A_{1(\mathsf{C}_{3v})} \uparrow O_h$ and $E_{(\mathsf{C}_{3v})} \uparrow O_h$ equal to 8 and 16, respectively.

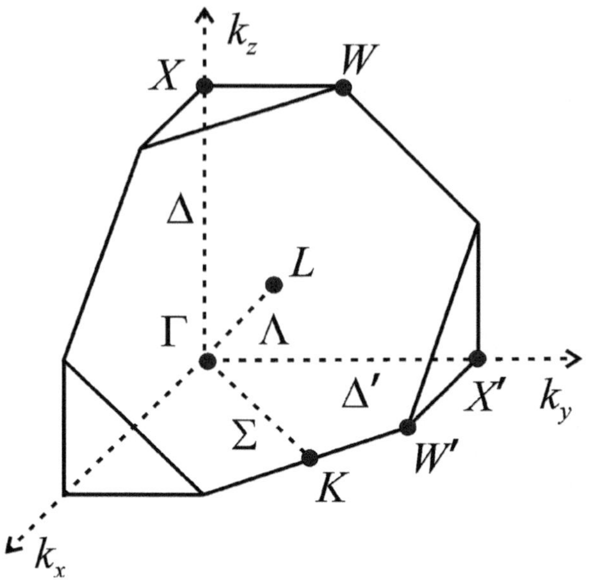

Fig. 3.9 1/8 part of a BZ for the face-centered cubic lattice

Table 3.9 Compatibility relations for IRs in the transition from the center of the zone Γ to the directions Λ and Δ

Group (k−point)	IRs									
$O_h\,(\Gamma)$	A_{1g}	A_{2g}	E_g	T_{1g}	T_{2g}	A_{1u}	A_{2u}	E_u	T_{1u}	T_{2u}
$C_{3v}\,(\Lambda)$	A_1	A_2	E	A_2+E	A_1+E	A_2	A_1	E	A_1+E	A_2+E
$C_{4v}\,(\Delta)$	A_1	B_1	A_1+B_1	A_2+E	B_2+E	A_2	B_2	A_2+B_2	A_1+E	B_1+E

Example 3.11 For a symmorphic space group O_h^5 with a face-centered cubic lattice, find the compatibility relations in the transition from the direction Δ (on the fourth-order axis) to the point $X' = b_3/2$ (on the fourth-order axis at the surface of the BZ) (Fig. 3.9).

In the direction Δ, the group of the wavevector k_Δ is the group C_{4v} and the star $\{k_\Delta\}$ consists of the six prongs (see Fig. 3.10a). The IRs of the space group are obtained by inducing the IRs $D^{\Delta,i}$ of the group $H_\Delta = C_{4v}$ into the full space group $G = O_h$. At the point X', the group of the wavevector k_X is the group D_{4h} and the star $\{k_X\}$ consists of three prongs (see Fig. 3.10b). IRs of the space group are obtained by inducing the IRs $D^{X,j}$ of the group $H_X = D_{4h}$ into the entire space group. Since $Ik_X = -kX = kX - b_3$ at point X, the wavevector group D_{4h} also includes the left coset of C_{4v} defined by inversion $D_{4h} = C_{4v} + IC_{4v}$. Thus making use of transitivity of induction theorem, we can write

$$D^{\Delta,i}_{(C_{4v})} \uparrow O_h = \left(D^{\Delta,i} \uparrow D_{4h}\right)_{(D_{4h})} \uparrow O_h$$

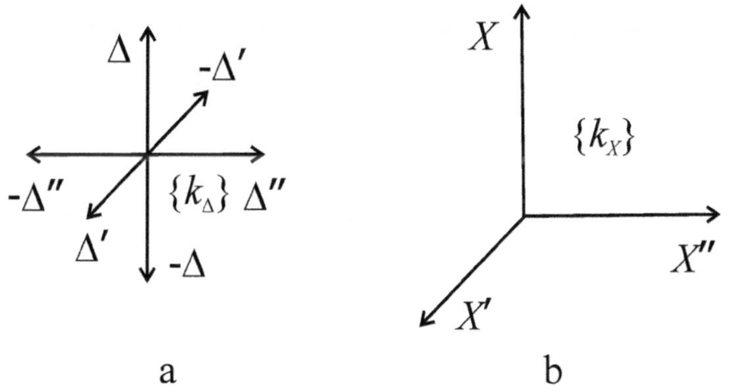

Fig. 3.10 Stars of the wavevector in a face-centered cubic lattice for k **a** in direction Δ and **b** point X

Table 3.10 Compatibility relations for direction Δ and point X for symmorphic space group O_h^5

Group ($k-$point)	IRs									
D_{4h} (X)	A_{1g}	A_{2g}	B_{1g}	B_{2g}	E_g	A_{1u}	A_{2u}	B_{1u}	B_{2u}	E_u
C_{4v} (Δ)	A_1	A_2	B_1	B_2	E	A_2	A_1	B_2	B_1	E

Here we use the symbols of the central extensions, i.e. point groups for the space group (its subgroups). As the wavevector k_Δ is approaching point X, the pairs of wavevectors $b_3/2 - \delta$ and $-b_3/2 + \delta + b_3$ merge ($\delta \longrightarrow 0$), resulting in the same phase factors for all block of the induced representation $D^{\Delta,i} \uparrow D_{4h}$. Thus, the representation $D^{\Delta,i} \uparrow D_{4h}$ becomes reducible and decomposes into IRs $D^{X,j}$. The decomposition can be done making use of Frobenius reciprocity theorem and one obtains the results presented in Table 3.10.

Note that the star of the wavevector in the direction Δ contains six rays, and the star at point X has only three rays, so the IRs in the direction Δ have twice the dimension (with the same dimension of small IRs) than at the point X. From Table 3.10, in particular, it follows that the IRs of the space group $(E)_{(C_{4v})} \uparrow O_h$ in the direction Δ decompose into two IRs $(E_u)_{(D_{4h})} \uparrow O_h$ and $\left(E_g\right)_{(D_{4h})} \uparrow O_h$ when passing from the direction Δ to the point X.

In non-symmorphic space groups in some points on the surface of a BZ projective, IRs differ from ordinary IRs of point groups. Therefore, compatibility relations when moving from a symmetrical direction to a more symmetrical point on the surface of the BZ may change. Consider a diamond structure—a space group O_h^7. At the center of a BZ in Γ point $s-$orbitals belong to IR A_{1g} and $p-$orbitals belong to IR T_{1u} and these orbitals don't interact. According to calculations T_{1u} is on the Fermi level and A_{1g} at the energy -25.8 eV [63]. At the Δ-direction T_{1u} level splits into E and A_1 and A_{1g} transfers into A_1 (see Table 3.9). Going to point X we must induce IRs of C_{4v} into D_{4h}, making use of formula for induction of projective IRs (3.1.11) (3.1.12).

Table 3.11 Part of characters' table of projective IRs t_i [60] at point X of a BZ for group O_h^7 and characters of IRs A_1 and E of group C_{4v}

	E	$2C_{4z}$	C_{2z}	$2\sigma_v$	$2\sigma_d$
t_1	2	0	−2	0	0
t_2	2	0	−2	0	0
t_3	2	0	2	0	2
t_4	2	0	2	0	−2
A_1	1	1	1	1	1
E	2	0	−2	0	0

Projective representations at point X are presented in Table 3.11. However, since the elements of C_{4v} don't change the $k-$vector at $X-$point, the part of factor system is trivial for the elements of C_{4v} (it follows from 3.11.11a) and (3.11.12)). Bearing in mind that Frobenius reciprocity theorem is valid for projective IRs also [57] we can easily decompose induced IRs making use of characters for subgroup C_{4v} presented in Table 3.10. Thus, we have $A_{1(C_{4v})} \uparrow D_{4h} = t_3$ and $E_{(C_{4v})} \uparrow D_{4h} = t_1 + t_2$. The former relation means that $s-$ and p_z-basis functions belong to the same projective IR of D_{4h} group and their energies coincide. This group-theoretical conclusion is confirmed by the band structure calculations, which result in merging of two bands, namely, $A_1(s)$ and $A_1(p_z)$, at point X merge into one band with the energy -14.3 eV [63]. Note that in the case of symmorphic space group (see Table 3.10) the A_1 in Δ-direction is compatible with A_{1g} and A_{2u} at point X. Since $s-$orbitals are even and p_z-orbitals are odd, they belong to different IRs, namely, A_{1g} and A_{2u} of D_{4h}. Thus, in symmorphic space group at point X, as in point group D_{4h}, $s-$ and p_z-orbitals belong to different IRs (A_{1g} and A_{2u}, respectively) and don't interact.

3.13 Additional Degeneracy in Crystals Due to Time-Reversal Symmetry

Consider a magnetic space group M of type- II, given by

$$M = G + \vartheta G \tag{3.13.1}$$

where ϑ is time reversal and a magnetic space group of type-III given by

$$M = H + A(G - H) \tag{3.13.2}$$

where A is antiunitary element, which may be a time reversal ϑ or the product ϑg_i, where g_i is an element $\{\alpha_i|\tau_i\}$. Since a translation subgroup T is not only an invariant subgroup of G but also an invariant subgroup of M, we can represent M as a left coset decomposition with respect to T:

$$M = \sum_i g_i T + \sum_j g'_j T, \tag{3.13.3}$$

where primed g'_j are antiunitary elements and unprimed g_i are unitary elements. All these elements g_i and g'_j form a factor group M/T.

The magnetic little co-group $\bar{Q}$ is the group of all unprimed elements $g_i = \{\alpha_i|\tau_i\}$, which belong to the wavevector co-group $\bar{H}$ (i.e. $\alpha_i k = k + b$) together with all primed elements $g'_j = \vartheta \left\{\alpha'_j|\tau'_j\right\}$, whose rotational parts α'_j transform k into $-k$ (i.e. $\alpha'_j k = -k + b$). Two cases occur, either no such α'_j exist at all and then $\bar{Q} = \bar{H}$, or there exist an equal number of primed and unprimed elements so that $\bar{H}$ is a subgroup of $\bar{Q}$ of index 2.

It should be noted that the definition of $\bar{Q}$ is not connected with the action of ϑ on the vector k itself. The relation

$$\vartheta k = -k \tag{3.13.4}$$

is connected with physical interpretation of k as quasi-momentum and is not used in the definition of magnetic little co-group [19].

The type of corepresentation is defined by the value of the following sum over the elements of $\bar{Q}$, which change k to $-k$.

$$\nu\left(\vartheta^2\right) \sum_{\bar{Q}-\bar{K}} \hat{\chi}\left[\left(\alpha'_j\right)^2\right] \omega_S\left(\alpha'_j, \alpha'_j\right) \omega_\tau\left(\alpha'_j, \alpha'_j\right) = \begin{cases} \left|\bar{H}\right| \text{ (a)} \\ -\left|\bar{H}\right| \text{ (b)} \\ 0 \text{ (c)} \end{cases} \tag{3.13.5}$$

where $\nu\left(\vartheta^2\right) = 1$ for integer spin and $\nu\left(\vartheta^2\right) = -1$ for half-integer spin. Also $\omega_S\left(\alpha'_j, \alpha'_j\right) = -1$ for half-integer spin, if the angle of the product $\left(\alpha'_j\right)^2 \geqslant 2\pi$, and $\omega_S\left(\alpha'_j, \alpha'_j\right) = 1$ in all other cases. The factor $\omega_\tau\left(\alpha'_j, \alpha'_j\right)$ appears in the surfaces of a BZ in non-symmorphic space groups. Note that $\left\{\alpha'_j|\tau'_j\right\}^2 \in H$ and $\hat{\chi}\left[\left(\alpha'_j\right)^2\right]$ is a character of projective representation of $\bar{H}$ and that all phase factors are written explicitly in (3.13.5). The product of two factors $\omega_S\left(\alpha'_j, \alpha'_j\right) \omega_\tau\left(\alpha'_j, \alpha'\right)$ may be estimated from the tables of space-group IRs [60].

Consider a magnetic grey space group, which can be written as $M = G(E + \vartheta)$, where G is a space group and ϑ is time reversal. It will be shown that the time-reversal symmetry in addition to Kramers' degeneracy, which connects states at k and $-k$, in some cases may cause degeneracy among small IRs $D^{k\kappa}$ of wavevector group H.

For a given wavevector k, the magnetic little co-group $\bar{Q}$ is written, depending on the group structure, in one of the three forms:

$$\bar{Q} = \bar{H} \tag{3.13.6}$$

$$\bar{Q} = \bar{H} + \vartheta \bar{H} \tag{3.13.7}$$

$$\bar{Q} = \bar{H} + A\bar{H} \tag{3.13.8}$$

where A is some antiunitary element which is a product of ϑ and some nontrivial element g' of the space group.

If $\bar{Q}$ is given by formula (3.13.6) there are no elements in G which transform k into $-k$, i.e. $-k$ doesn't appear in the star $\{k\}$ and the set $\bar{Q} - \bar{H}$ in (3.13.5) is void and additional degeneracy at $\bar{k}$ is absent. In this case, the IR $D^{k\kappa} \uparrow G(g)$ of a group G is constructed by the induction. ICRs of the whole group M are constructed from the IRs of the whole group G according to case (c):

$$\Delta(g) = \begin{pmatrix} D^{k\kappa} \uparrow G(g) & 0 \\ 0 & D^{k\kappa *} \uparrow G(g) \end{pmatrix} \tag{3.13.9}$$

$$\Delta(\vartheta g) = \begin{pmatrix} 0 & \nu D^{k\kappa} \uparrow G(g) \\ D^{k\kappa *} \uparrow G(g) & 0 \end{pmatrix} \tag{3.13.10}$$

where $\nu = 1$ for integer spin and $\nu = -1$ for half-integer spin. Since $exp(ikt)^* = exp(-ikt)$, $D^{k\kappa *}$ is a small IR of little group of wavevector $-k$. In this case, $+k$ and $-k$ are not in the same star and eigenvalues at $+k$ and $-k$ are degenerated due to time-reversal symmetry (Kramers' degeneration)

If $\bar{Q}$ is given by formula (3.13.7), the elements of $\bar{H}$ themselves transform k into $-k + b$, where b is a reciprocal lattice vector. This takes place at symmetry points for which k is zero or half a reciprocal lattice vector.

If $\bar{Q}$ is given by formula (3.13.8), $-k$ appears in the star of k but is not equivalent to it. The cases (3.13.7) and (3.13.8) can be considered together as far as degeneracies are concerned. Because the spectra of eigenvalues at k and $-k$ are already identical when only the symmetry of G is considered, the only additional degeneracy that can arise because of the presence of the additional elements ϑG in M will involve extra degeneracies among the various eigenvalues at k.

Therefore, the extra degeneracies, arising due to the inclusion of time-reversal symmetry in the case when $-k$ is in the star of k we have to determine whether the corepresentations of $\bar{Q}$ belong to case (a), case (b) or case (c). In case (a), there is no increase in degeneracy at k, but in cases (b) and (c) a doubling of the degeneracy occurs at k. In the case (b) the extra degeneracy is between two sets of eigenvalues both belonging to the same IR.

Example 3.12 Find the types of ICRs for the space groups D_{6h}^1 and D_{6h}^4 for the $k-$points on the upper face of a BZ, i.e. $k = \mu b_1 + \nu b_2 + b_3/2$. These groups have the same rotational elements and BZs, which are regular hexagonal prisms and wavevectors considered are on upper face beyond symmetrical points. The elements of wavevector group $\bar{H}$ and of magnetic little co-group $\bar{Q}$ can be easily found. These elements are presented in Table 3.12 together with their translations.

There are two terms in the sum (3.13.5).

The first term in the sum (3.13.5) is written as

$$\nu\left(\vartheta^2\right)\{I|0\}\{E|0\}\{I|0\}\{E|0\} = \nu\left(\vartheta^2\right)\{E|0\} = \begin{cases} 1, & \text{for } t_i(D_{6h}^1), \text{ and } t_i(D_{6h}^4) \\ -1 & \text{for } p_i\, p_i(D_{6h}^1), \text{ and } p_i(D_{6h}^4) \end{cases}$$

and resulting factor equals to 1 for both types of IRs. For the second term, we have

$$\nu(\theta^2)\{I|0\}\{\sigma_z|\tau_z\}\{I|0\}\{\sigma_z|\tau_z\} = \nu(\theta^2)\{C_{2z}|-\tau_z\}\{C_{2z}|-\tau_z\} = \nu(\theta^2)$$
$$\{E|-2\tau_z\}\,\omega(C_{2z}, C_{2z}) =$$
$$= \begin{cases} 1, & \text{for } t_i(D_{6h}^1), \text{ and } p_i(D_{6h}^4) \\ -1 & \text{for } t_i(D_{6h}^4), \text{ and } p_i(D_{6h}^1) \end{cases}$$

This result can be obtained bearing in mind that $\chi\left(\{E|-2\tau_z\}\right) = -1$ for $k = \mu b_1 + \nu b_2 + b_3/2$ for t_i and p_i in D_{6h}^4 symmetry. Also $\omega(C_{2z}, C_{2z}) = -1$ and $\nu(\theta^2) = -1$ for double-valued IRs.

Finally, for the group D_{6h}^4 we obtain that the sum (3.13.5) equals 0 for single-valued IRs and equals -2 for double-valued IR, what corresponds to cases (c) and (b), respectively

Similarly for group D_{6h}^1 the sums equal to 2 and 0 for single-valued and double-valued IRs and correspond to cases (a) and (c), respectively. All cases are presented in Table 3.13.

Table 3.12 Elements of magnetic little co-group for space groups D_{6h}^1 and D_{6h}^4 and $k = \mu b_1 + \nu b_2 + b_3/2$ where $\tau_z = b_3/2$ [60]

	$\bar{H}$		$\bar{Q} - \bar{H}$	
	E	σ_z	I	C_{2z}
D_{6h}^4	(000)	$(00\tau_z)$	(000)	$(00\tau_z)$
D_{6h}^1	(000)	(000)	(000)	(000)

Table 3.13 IRs and ICRs of little group C_s (group of the wavevector $k = \mu b_1 + \nu b_2 + b_3/2$) for the space groups D_{6h}^4 and D_{6h}^1 (t_i—single-valued IRs, p_i—double-valued IRs) [60]

IR, D_{6h}^4	E	σ_z	Case, ICR	IR, D_{6h}^1,	E	σ_z	Case, ICR
$\hat{t}_1$	1	i	(c)	$\hat{t}_1$	1	1	(a)
$\hat{t}_2$	1	$-i$	$\hat{t}_1 + \hat{t}_2$	$\hat{t}_2$	1	-1	t_1, t_2
$\hat{p}_1$	1	1	(b)	$\hat{p}_1$	1	i	(c)
$\hat{p}_2$	1	-1	$2\hat{p}_1, 2\hat{p}_2$	$\hat{p}_2$	1	$-i$	$p_1 + p_2$

3.14 Time-Reversal Symmetry and Band Sticking in Non-symmorphic Space Groups

In a strong spin-orbital coupling case in low-symmetry points, spin-up and spin-down bands split, but such states may merge in higher symmetry points or due to time-reversal symmetry. In this section, we consider this effect taking as an example non-symmorphic space group D_2^4 $(P2_12_12_1)$ whose elements are as follows:

$$E, \{C_{2x}|1/2, 1/2, 0\}, \left\{C_{2y}|0, 1/2, 1/2\right\}, \{C_{2z}|1/2, 0, 1/2\} \tag{3.14.1}$$

A basis domain of a BZ for the group D_2^4 is shown in Fig. 3.11.

We consider three directions: Σ $(k_x, 0, 0)$, $A(k_x, 0, b_3/2)$ and $E(k_x, b_2/2, b_3/2)$ and points $k_x = 0$ and $k_x = b_1/2$ in these directions (see Fig. 3.11). Wavevector groups are the same in all these directions and also in all points, but the types of corepresentations differ due to the non-symmorphic structure of the space group. The small double-valued IRs [60] for k in these directions are presented in Table 3.14. The subscripts of capital letter P_i correspond to the set of IRs for any particular group and $k-$ vector and subscript of small $\hat{p}_i$ corresponds to the number of IR in this set. The IRs on lines Σ and A are projective, but on line E spin and spatial factors equal to -1 and their product equals to 1 resulting in ordinary IR. In this case, the difference from spinless case is due to the factor $\nu\left(\vartheta^2\right)$.

In all these directions, C_{2x} belongs to a wavevector group and the elements C_{2y} and C_{2z} change k to $-k$. Hence, it follows that the magnetic little co-group $\bar{Q}$ consists of the elements ϑC_{2y} and ϑC_{2z} in addition to the elements of a little co-group $\bar{H}$. In such a case, the space-group elements not included in $\bar{H}$ will be used for the calculations of a sum in Herring criterion. For the group under consideration, we can write the sum in Herring criterion as

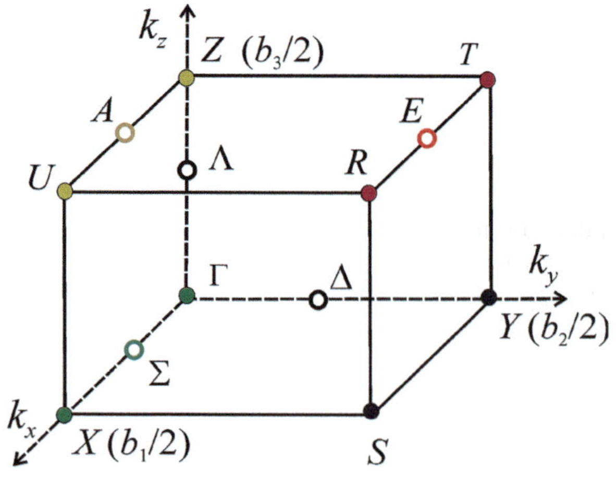

Fig. 3.11 Basis domain of a BZ for the space group D_2^4 $(P2_12_12_1)$

Table 3.14 Projective double-valued IRs of the wavevector group in directions Σ, A and E for the space group D_2^4 $(P2_12_12_1)$. The subscript of capital P corresponds to the set of IRs and subscript of small $\hat{p}$ is the number of IR in this set [60]

P_{15} Σ, A			P_{20} E		
	E	C_{2x}		E	C_{2x}
$\hat{p}_1$	1	i	$\hat{p}_1$	1	1
$\hat{p}_2$	1	$-i$	$\hat{p}_2$	1	-1

Table 3.15 Factors ν, ω_s and ω_τ required to calculate sum in Herring criterion (3.13.5) on the symmetry line Σ, A and E in BZ for group D_2^4 $(P2_12_12_1)$ (see Fig. 3.11), the value of the sum (3.13.5) and type of corepresentation

	ν	ω_s	ω_τ		Sum	Type
Elements	All	All	ϑC_{2y}	ϑC_{2z}		
Σ $(k_x, 0, 0)$	-1	-1	1	1	2	(a)
A $(k_x, 0, b_3/2)$	-1	-1	1	-1	0	(c)
E $(k_x, b_2/2, b_3/2)$	-1	-1	-1	-1	-2	(b)

$$\sum_{\bar{Q}-\bar{H}} = \chi\left(\vartheta\left\{C_{2y}|\tau_1\right\}\right)^2 + \chi\left(\vartheta\left\{C_{2z}|\tau_2\right\}\right)^2 = \nu\left(\vartheta^2\right)\left(\omega_{s1}\omega_{\tau1} + \omega_{s2}\omega_{\tau2}\right)\chi\left(\{E|0\}\right) \tag{3.14.2}$$

where $\nu\left(\vartheta^2\right) = -1$ for double-valued IRs and since the squares of rotational parts result in 2π, all $\omega_s = -1$. The squares of space-group elements in 3.14.2 are written as

$$\left\{C_{2y}|0, 1/2, 1/2\right\}\left\{C_{2y}|0, 1/2, 1/2\right\} = \{E|0, 1, 0\} \tag{3.14.3}$$

$$\{C_{2z}1/2, 0, 1/2\}\{C_{2z}1/2, 0, 1/2\} = \{E|0, 0, 1\} \tag{3.14.4}$$

Corresponding phase factors $\omega_{\tau i}$ depend on the wavevector k_s and are obtained from the standard relation:

$$\varpi_{\tau i} = \exp(-ik_s \cdot \tau_i\ ) \tag{3.14.5}$$

All phase factors, the sum in Herring criterion (3.13.5) and type of corepresentations on lines Σ, A and E are presented in Table 3.15. It is seen in Table 3.15 that in the case of the same groups $\bar{Q}$ and $\bar{H}$ the types of corepresentations depend on improper translations and on the wavevector.

Going over to the symmetry points we see that $\bar{Q} = \bar{H}$ and in Herring criterion the squares of the elements of $\bar{H}$ should be calculated. This can be done using tabular projective representations [60]. Projective IRs of wavevector groups at points Γ and X are presented in Table 3.16.

Table 3.16 Projective double-valued IRs of the wavevector group D_2 in points Γ and X, projective factors $\nu\ \omega_\tau\omega_s$ (3.13.5), sums in Herring criterion and type of corepresentations. The subscript of capital P corresponds to the set of IRs and subscript of small $\hat{p}$ is the number of IR in this set [60]

	E	C_{2x}	C_{2y}	C_{2z}	Sum	Type
E' [1)], $\Gamma(0,0,0)$	2	0	0	0	4	(a)
$\nu\omega_\tau\omega_s$	-1	1	1	1		
P_{25}, $X(b_1/2,0,0)$						
$\hat{p}_1$	1	i	i	1	0	(c)
$\hat{p}_2$	1	i	$-i$	-1	0	(c)
$\hat{p}_3$	1	$-i$	i	-1	0	(c)
$\hat{p}_4$	1	$-i$	$-i$	1	0	(c)
$\nu\omega_\tau\omega_s$	-1	1	1	-1		

$\longrightarrow$

[1)]Double-valued IRs of D_2 are presented in Sect. 1.8

At point Γ we have double-valued IR E' of group D_2, considered in Chap. 1. Bearing in mind that squares of the elements C_{2i} $(i = x, y, z)$ result in rotation by 2π, we get

$\nu\left(\vartheta^2\right)\sum_{\vartheta h} = (-1)\,(2-2-2-2) = 4$

Consequently at point Γ we have the case (a) and there is no additional degeneracy due to time reversal. Nevertheless the two spin states are degenerated due to spatial symmetry.

At high-symmetry points the product of the factors $\omega_\tau\omega_s$ can be estimated from the relation:

$$\hat{D}(C_{2i})\hat{D}\,(C_{2i}) = \omega_\tau\omega_s\hat{D}(E) \tag{3.14.6}$$

It is seen in Table 3.16 that at point $X(b_1/2,0,0)$ these factors equal to -1 for C_{2x} and C_{2y} and we have

$\nu\sum_{\vartheta h} = (-1)\,(1-1-1+1) = 0$

Thus corepresentations at point $X(b_1/2,0,0)$ belong to type (c). In the case (c), two complex conjugated IRs are sticked into one corepresentation. In order to define which representations stick in one corepresentation the projective representations $\hat{p}_i$ should be transformed into space group IRs p_i by multiplying by the factors $\exp(-ik\cdot\tau)$. Space group IRs for point X are presented in Table 3.17, where we see that ICRs are $p_1 + p_2$ and $p_3 + p_4$.

Projective IRs and projective factors for points Z and U are presented in Table 3.18. Making use of these factors Herring criterion is easily estimated and we obtain types (c) and (b) for points Z and U, respectively. In order to find out which of the representations are combined space-group representations are constructed in Table 3.19. Since only C_{2y} and C_{2z} have improper translations in $z-$direction these projective representations are multiplied by $-i$. Thus we obtain that at point Z in

Table 3.17 IRs of space group D_2^4 at point $X(b_1/2, 0, 0)$

$P_{25} \times \exp(-ik \cdot \tau_i)$	E	C_{2x}	C_{2y}	C_{2z}
p_1	1	1	i	$-i$
p_2	1	1	$-i$	i
p_3	1	-1	i	i
p_4	1	-1	$-i$	$-i$

Table 3.18 Projective double-valued IRs of the wavevector group D_2 in points Z and U, projective factors $\omega_\tau \omega_s$ (3.13.5), sums in Herring criterion and type of corepresentations. The subscript of capital P corresponds to the set of IRs and subscript of small $\hat{p}$ is the number of IR in this set [60]

	E	C_{2x}	C_{2y}	C_{2z}	Sum	Type
P_{19} $Z(0, 0, b_3/2)$						
$\hat{p}_1$	1	i	1	i		
$\hat{p}_2$	1	i	-1	$-i$		
$\hat{p}_3$	1	$-i$	1	$-i$		
$\hat{p}_4$	1	$-i$	-1	i		
$\nu\omega_s\omega_\tau$	-1	1	-1	1	0	(c)
P_{27} $U(b_1/2, 0, b_3/2)$						
$\hat{p}_1$	$\begin{pmatrix} 1 & 0 \\ 0 & 1 \end{pmatrix}$	$\begin{pmatrix} i & 0 \\ 0 & -i \end{pmatrix}$	$\begin{pmatrix} 0 & 1 \\ 1 & 0 \end{pmatrix}$	$\begin{pmatrix} 0 & i \\ -i & 0 \end{pmatrix}$		
$\nu\omega_s\omega_\tau$	-1	1	-1	-1	-4	(b)

Table 3.19 IRs of space group D_2^4 at point $Z(b_1/2, 0, 0)$

P_{19}	E	C_{2x}	C_{2y}	C_{2z}
p_1	1	i	$-i$	1
p_2	1	i	i	-1
p_3	1	$-i$	$-i$	-1
p_4	1	$-i$	i	1

corepresentations are sticked IRs $p_1 + p_4$ and $p_2 + p_3$. At point U the case (b) occurs and the representation appears twice.

3.15 Local Symmetry and Band Representations in Crystals

In this section, we consider a connection between Bloch functions and symmetry-adapted localized orbitals (Wannier functions). A general lattice associated with a space group G is the set of all the distinct points A_i which can be obtained from a given point A_1 in a space, by transformations g_i belonging to G. If L is a local symmetry group of an atom A_1 then the symmetry group of A_i is $g_i L g_i^{-1}$. All the points A_i of a lattice are equivalent and can be used to generate the whole lattice. Thus we see that the term lattice which is used in a solid-state theory is equivalent to a group-theoretical term orbit. The space group G leaves the Hamiltonian $\hat{H}$ invariant. The eigenfunctions of $\hat{H}$ can be constructed from the basis functions of the local Hamiltonians $\hat{H}_L$ on all points of lattice. From the set of localized orbitals which is invariant under all the operations of the space group, it is possible to construct basis functions for all $k-$vectors in a BZ, forming a band, and this approach is called band representations. The aim of the present section is to consider symmetry properties of such basis sets. The construction of orbitals in a space-group symmetry on the basis of atomic orbitals may be accomplished in a similar way as for molecules in a point group symmetry, i.e. by induction from the local point group of an atom into the entire symmetry group. Since the number of atoms in a crystal is infinite, this technique results in an infinite number of space-group IRs, forming a band, originating from an orbit of atoms. Since elementary cell of a solid contains a set of atoms, the induction of wavefunctions of all non-equivalent atoms results in for each k-vector a set of equivalent IRs, whose functions interact, thus forming an actual band structure.

Consider one orbit of atoms in a non-symmorphic space group G with elements $g_{it} = \{\alpha_i|\tau_i + t\}$, where α_i is a point, group element, τ_i is its improper translation and t is an element of translation subgroup T of dimension N. This notation corresponds to the left coset decomposition of the space group with respect to invariant subgroup of translations:

$$G = \sum_i g_i T, \; g_i = \{\alpha_i|\tau_i\} \tag{3.15.7}$$

where the elements g_i form a factor group $\bar{G}$. If atom A_1 is chosen as the origin of coordinate system, then the local symmetry group L of atom A_1 consists of all symmorphic elements $g_l = \{\alpha_l|0\}$. The factor group is represented as a left coset decomposition relative to L:

$$\bar{G} = \sum_{\lambda=1}^{p} g_\lambda L, \; p = |\bar{G}| / |L| \tag{3.15.8}$$

where $g_\lambda = \{\alpha_\lambda|\tau_\lambda\}$ are non-symmorphic elements, transforming atom A_1 to equivalent positions A_λ in the same primitive sell. Under the action of the elements of local group L atomic wavefunctions are transformed by the IRs B^j of this group. Consider one of such IRs B and its basis $\{\psi_B\}$ of dimension l. When acting by all

g_λ on the wavefunctions $\{\psi_B\}$ we obtain functions on all equivalent centers A_λ of a primitive cell, whose local symmetry groups are $L^\lambda = g_\lambda L g_\lambda^{-1}$. We will call this set a suborbit of atom A_1. When acting by the elements of translation group T on the basis set of a suborbit we obtain the basis set of induced representation $B_{(L)} \uparrow G$ whose dimension is lpN. Since the number of elements N of translation group is infinite this IR in a general case is decomposed into all IRs $D_{(H)}^{k\kappa} \uparrow G$ of the space group G, where k runs over the basis domain of a BZ and κ runs over all small IRs of a little group H.

Our task is to establish relations between these two types of induced representations, i.e. reducible, induced from local group $B_{(L)} \uparrow G$ and irreducible, induced from the wavevector group $D_{(H)}^{k\kappa} \uparrow G$. It can be done using Mackey theorem on the intertwining number of induced representations. To do this, we decompose the space group into double cosets relative to subgroups H and L:

$$G = \sum_{\sigma} H d_\sigma L \tag{3.15.9}$$

For every double coset σ, we define an intersection group $M_\sigma = H \cap d_\sigma L d_\sigma^{-1}$. Also subduced representations $D^{k\kappa} \downarrow M_\sigma$ and $B^\sigma \downarrow M_\sigma = B(d_\sigma^{-1} g d_\sigma) \downarrow M_\sigma$ are introduced. Then the intertwining number $I\left(B_{(L)} \uparrow G / D_{(H)}^{k\kappa} \uparrow G\right)$ (the frequency of appearance of IR $D_{(H)}^{k\kappa} \uparrow G$ in the decomposition of $B_{(L)} \uparrow G$) is given by the following formula:

$$I\left(B_{(L)} \uparrow G / D_{(H)}^{k\kappa} \uparrow G\right) = \sum_{\sigma} I\left(B^\sigma \downarrow M_\sigma / D^{k\kappa} \downarrow M_\sigma\right) \tag{3.15.10}$$

The intertwining number for each double coset σ in the right-hand side of (3.15.10) is calculated as

$$I\left(B^\sigma \downarrow L_\sigma / D^{k\kappa} \downarrow L_\sigma\right) = \frac{1}{|M_\sigma|} \sum_{m \subset M\sigma} \chi(B^\sigma(m)) \chi^*(D^{k\kappa}(m)) \tag{3.15.11}$$

To construct a basis set, corresponding to wavevector group H, the B^σ is induced into H, and then $B^\sigma \uparrow H$ is decomposed into IRs $D^{k\kappa}$ of H. Further induction of $D^{k\kappa}$ into G results in IRs $D^{k\kappa} \uparrow G$ of the space group. It follows from Mackey intertwining number theorem that distributions of localized orbitals by the prongs of the star $\{k\}$ differ for symmorphic and non-symmorphic groups. If the group is symmorphic, the local symmetry group L coincides with the central extension of G. In this case, suborbit of atom A_1 consists of one element and there is only one double coset in (3.15.9). To construct basis set one should project local wavefunctions of the atom A_1 on IRs $D^{k\kappa}$ and then induce the result into the entire group. Thus all atomic wavefunctions will be distributed over all prongs of the star$\{k\}$. Consider a double coset defined by identity element. When projecting the wavefunctions of

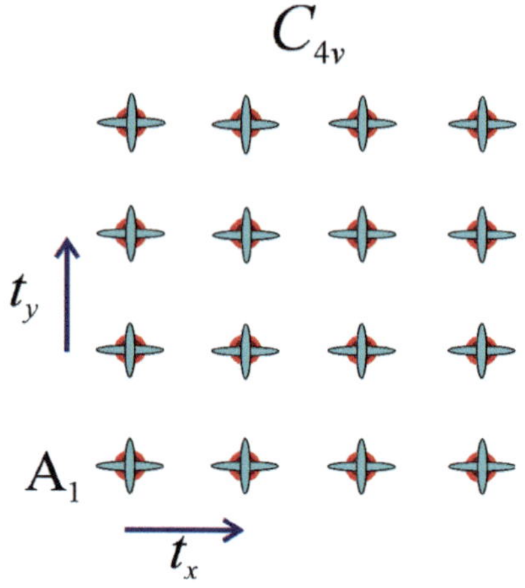

Fig. 3.12 Square lattice with p_x- and p_y-orbitals on each atomic center

atom A_1, by the elements of intersection $M = H \cap L$ we obtain basis functions $D^{k\kappa}$ corresponding to the prong k of the star $\{k\}$. When acting by the elements $g_\lambda \notin H$ in (3.15.8) we project the functions of atom $g_\lambda A_1$ on the prong $k_\lambda = g_\lambda k$ of the star $\{k\}$. For non-symmorphic space groups there are two cases. If the element g_λ belongs to the group H, then the atomic basis set $g_\lambda A_1$ is projected on the prong k. In opposite case, i.e. if $g_\lambda \notin H$, this atomic basis set belongs to k_λ distinguished from k. These cases are illustrated by examples below.

Example 3.13 Basis functions for two-dimensional square lattice (C_{4v} symmetry group) with p_x- and p_y-orbitals. Consider a square lattice of C_{4v} symmetry, defined by two equal and orthogonal vectors t_x and t_y with basis functions (Wannier functions) p_x and p_y on center A_1 (see Fig. 3.12). The basis functions on other centers are obtained by lattice translations $R_i = n_i t_x + m_i t_y$.

The wavevector star for k_1 a general point in a basis domain of a BZ is shown in Fig. 3.13a. Basis domain is a sector $0 < \theta \leq \pi/4$. The wavevector group at k_1 consists of identity element E only and there are two basis sets, originating from p_x and p_y orbitals. Basis functions for other prongs are obtained by the action of point group elements (see Table 1.1).

If the initial vector k_1 is in symmetry plane (see Fig. 3.13b) its symmetry group has two IRs A and A' whose basis functions are $p_x + p_y$ and $p_x - p_y$, respectively. The star of this wavevector consists of four prongs. When acting by projection operators one obtains basis functions of all prongs, presented in Table 3.20.

Example 3.14 Basis set constructed from atomic $s-$orbitals in Δ- and Λ-directions in diamond crystal structure. Diamond crystal structure belongs to the space group O_h^7 consisting of one orbit of carbon atoms. Diamond structure is illustrated in Fig. 3.14. The primitive unit cell consists of two atoms with local symmetry of both

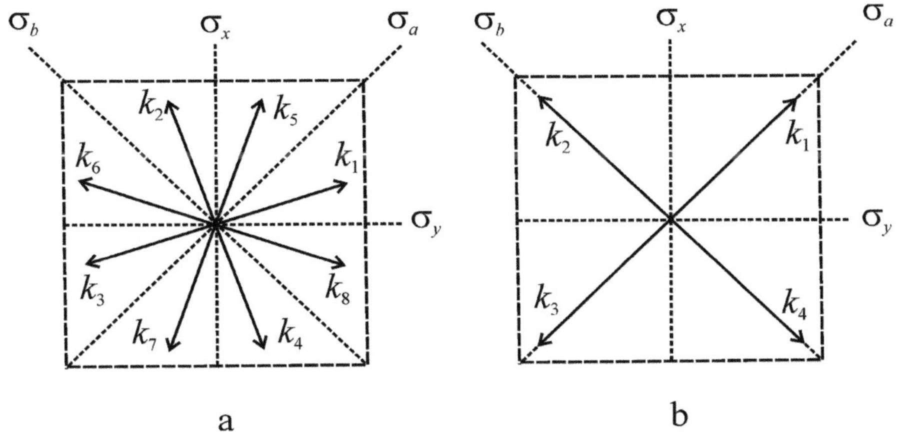

Fig. 3.13 Wavevector stars in a square lattice: **a** k_1 in a general point, **b** k_1 in diagonal of a square

Table 3.20 Basis functions for the prongs of $k-$vector stars shown in Fig. 3.13 a. b. built from p_x- and p_y-orbitals of atom A_1 in square lattice of C_{4v} symmetry (see Fig. 3.12). Local basis functions on other atomic centers $A_i = R_i A_1$ appear with the phase factors $exp(-ik_j \cdot R_i)$

		Wavevector group H			
k_j	g	E (a)		C_s (b)	
k_1	E	p_x	p_y	$p_x + p_y$	$p_x - p_y$
k_2	C_4	p_y	$-p_x$	$-p_y + p_x$	$p_y + p_x$
k_3	C_2	$-p_x$	$-p_y$	$-p_x - p_y$	$-p_x + p_y$
k_4	C_4^3	$-p_y$	p_x	$p_x - p_y$	$p_x + p_y$
k_5	σ_a	p_y	p_x		
k_6	σ_x	$-p_x$	p_y		
k_7	σ_b	$-p_y$	$-p_x$		
k_8	σ_y	p_x	$-p_y$		

atoms T_d. Space inversion appears with improper translations $\{I|t/4, t/4, t/4\}$. It is clear from Fig. 3.14 that if the center of coordinate system is chosen at the atom A_1, the element $\{I|t/4, t/4, t/4\}$ transforms A_1 into A_2 and also A_2 into A_1.

Little groups in directions Δ and Λ are the groups C_{4v} and C_{3v}, respectively (see Fig. 3.9). Let's first consider the direction Δ. The double coset decomposition (3.15.9) of O_h relative to C_{4v} and T_d consists of one double coset defined by identity element only. The intersection group $M = H \cap L$ is a group C_{2v}. The s_1-orbital on atom A_1 belongs to IR A_1 of group C_{2v}. Element C_4 of the left coset doesn't belong to T_d and transforms s_1 into s_2. The induction from C_{2v} into C_{4v} results in IRs A_1 and B_2 of group C_{4v}. Thus functions s_1 and s_2 of two atomic centers belong to one IR A_1 (or B_2) of the wavevector group C_{4v}. These functions are presented in Table 3.21 and are shown in Fig. 3.15a. Linear combination A_1 is bonding, since the

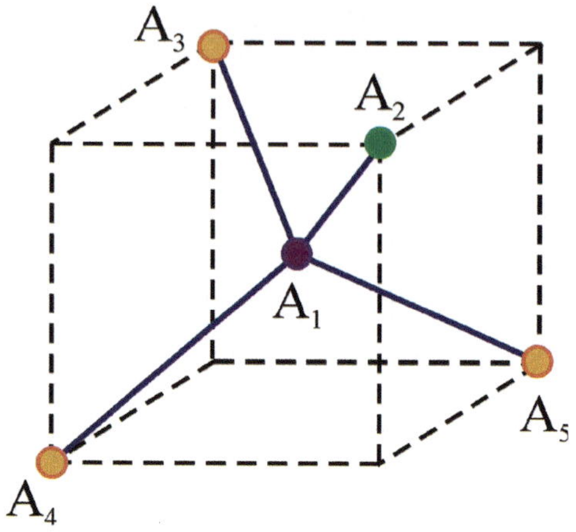

Fig. 3.14 Diamond crystal structure

Table 3.21 Basis sets constructed from s−orbitals of atoms A_1 and A_2 in diamond structure for the wavevector stars $\{\Delta\}$ and $\{\Lambda\}$ (Figs. 3.9 and 3.15). In the direction A, there are two independent basis sets denoted a and b

Direction Δ			Direction Λ		
Prongs k_i	Small IR, basis		Prongs k_i	Small IR, basis	
	A_1	B_2		A_1	A_1
$\Delta_1 - \Delta_6$	$s_1 + s_2$	$s_1 - s_2$	$a : \Lambda_1, \Lambda_3, \Lambda_6, \Lambda_8$	s_1	s_2
			$b : \Lambda_2, \Lambda_4, \Lambda_5, \Lambda_7$	s_2	s_1

signs of s_1 and s_2 are the same and linear combination B_2 is antibonding, since the signs are opposite. Thus, if there is only one double coset in the decomposition of the space group relative to local symmetry group L and wavevector group H, then the bonding and antibonding orbitals have different labels of IRs of the space group, namely, $A_{1(C_{4v})} \uparrow G$ and $B_{2(C_{4v})} \uparrow G$.

In the direction Λ little group C_{3v} is a subgroup of T_d and there are two double cosets in (3.15.9), defined by identity element E and space inversion I. For $d_\sigma = E$ the intersection group $M_\sigma = H \cap L$ is a group C_{3v}. The elements of the wavevector group C_{3v} leave A_1 in place and element I moves A_1 into A_2. Thus the s_1 appears at the prongs Λ_i connected with prong Λ_1 by the elements of local group T_d. The left coset decomposition of T_d with respect to C_{3v} consists of C_2 rotations about coordinate axis. Thus s_1 appears at four prongs (see Fig. 3.15b). Since $\{I|t/4, t/4, t/4\}$ (or $\{C_4|t/4, t/4, t/4\}$) transfers A_1 to A_2 and these elements don't belong to the wavevector group C_{3v}, orbitals s_1 and s_2 appear at prongs connected by space inversion I or rotation C_4 (see Table 3.21 and Fig. 3.15b). There is one more basis set which corresponds to double coset defined by I. In this case, the projection of basis on the prong Λ_1 starts from the atom A_2 and function s_2 and distribution of s_1 and s_2 over prongs is opposite (see Table 3.21). These two invariant basis sets have the same

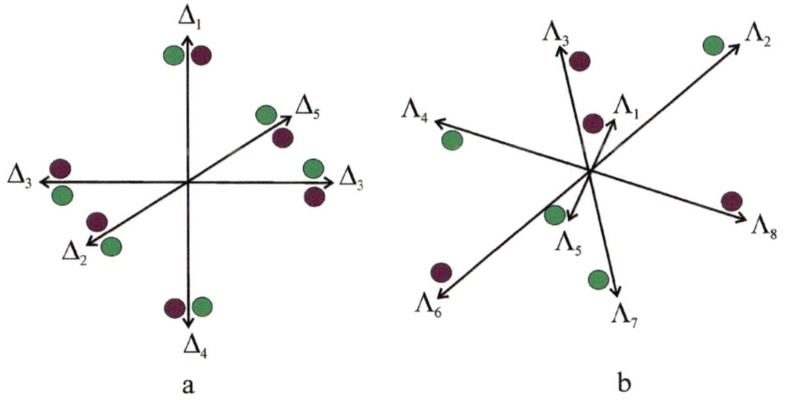

Fig. 3.15 Invariant basis sets constructed from the s-orbitals of atoms A_1 and A_2 (green and purple circles) for diamond crystal structure. a—direction Δ in a BZ, functions of both atoms contribute to all wavevectors and b—direction Λ in a BZ, functions of atoms A_1 and A_2 contribute to the wavevectors connected with the space inversion

symmetry and are able to interact, resulting in bonding or antibonding combinations (with plus or minus sign). However in contrast to the case of direction Δ, in the direction Λ bonding and antibonding orbitals belong to the same IR $A_{1(C_{3v})} \uparrow G$ of the space group.

3.16 Selection Rules for Electron Transitions in Solids

A transition between initial state belonging to IR Γ^i, and final state belonging to IR Γ^f caused by a perturbation potential belonging to IR Γ^p of the symmetry group G is forbidden by the symmetry if the inner Kronecker product $\Gamma^i \times \Gamma^p$ doesn't contain Γ^f, or equivalently the triple product $\Gamma^i \times \Gamma^p \times \Gamma^{f*}$ doesn't contain totally symmetric IR of G. In point group symmetry, this problem is easily solved by making use of existing tabled of point group IRs [2, 4, 11]. However in crystal solids the characters are presented for small IRs, and the entire IRs of the space groups are obtained by induction [19, 60]. C.J. Bradley [37] was the first who used the Mackey theorem for expanding Kronecker products of space-group IRs. In the present section, Bradley's technique is considered and an example is presented.

Electrons of valence bands of solids cannot be considered localized and for their description, IRs of space groups are used, which are characterized by the wavevector k (quasi-momentum) and a small IR $D^{k\kappa}$. Since the photon momentum in optical or UV region is much less than the electron momentum in solid, electronic transitions with absorption or emitting of photons are possible between two electrons with the same k-vectors. The selection rules are easy to find using small IRs of wavevector group H for electron states and for the dipole operator. Optical absorption spectra of solids make it possible to obtain the band gap. However, in many cases, the highest energy state of the valence band and the lowest energy state of the conduction band have different wavevectors and pure dipole transitions between such states are forbidden. However, optical transitions with the participation of phonons called indirect transitions are possible. In this case, the possible IRs of the transition operator are obtained as a product of the dipole operator and the IR of the phonon. For such a product of IR of a point group Γ^q (for the dipole operator) and induced IR $D^{\kappa}_{(H)} \uparrow G$ (for the phonon) following theorem is valid:

$$\Gamma^q \times \left(D^{\kappa}_{(H)} \uparrow G\right) = \left[\left(\Gamma^q \downarrow H\right) \times D^{\kappa}\right]_{(H)} \uparrow G \tag{3.16.1}$$

This theorem is a consequence of Mackey theorem [35] considered below. It follows from this relation that in order to find IRs in the decomposition of a direct product (3.16.1), it is necessary to expand the product $(\Gamma^q \downarrow H) \times D^{\kappa}$ into the IRs of the wavevector group H, and then induce the result into the entire space group. In particular, in the case of a cubic group, Γ^q is the IR T_{1u}.and in D_{6h} symmetry Γ^q is the sum of A_{2u} and E_{1u}.

The phonon symmetry corresponds to the IR of the space group with a nonzero wavevector. Therefore, to solve the problem of finding selection rules in crystals, a method is needed to decompose the product of the IRs of space groups into irreducible ones. Since the direct expansion in terms of orthogonality relations is rather complicated for the space groups, the method based on Mackey theorem is used.

Consider Mackey theorem for the direct product of induced representations [35, 37]. Let D be the IR of the group H of the wavevector k_1 and B be the IR of the group F of the wavevector k_2. The space group may be decomposed into left cosets relative to its subgroups H and F:

$$G = \sum_{\alpha=1}^{n_1} p_\alpha H, n_1 = |G| / |H| \tag{3.16.2}$$

$$G = \sum_{\beta=1}^{n_2} p_\beta F, n_2 = |G| / |F| \tag{3.16.3}$$

The actions of left coset representative in (3.16.2) and (3.16.3) on vectors k_1 and k_2 result in the stars of these vectors $\{k_1\}$ and $\{k_2\}$, consisting of n_1 and n_2 prongs, respectively.

Let us decompose the space group G into double cosets with respect to its subgroups H and F:

$$G = \sum_{\delta} H d_\delta F \tag{3.16.4}$$

Since the subgroup of translations is included in the wavevector groups, the representatives of the double cosets d_δ don't include lattice translations $t \in T$. For each d_δ, including the identity element, consider the intersection subgroup:

$$M_\delta = H \cap d_\delta F d_\delta^{-1} \tag{3.16.5}$$

and the representation of this group, which in general case is reducible:

$$P^\delta(m) = D(m) \times B(d_\delta^{-1} m d_\delta) \tag{3.16.6}$$

Then, according to Mackey theorem [19, 35, 37], the direct product of the induced representations is given by the following relation:

$$\left(D_{(H)} \uparrow G\right) \times \left(B_{(F)} \uparrow G\right) = \sum_\delta P^\delta_{(M_\delta)} \uparrow G \tag{3.16.7}$$

Mackey theorem admits a simple geometric interpretation, by construction of the direct product of wavevector stars and decomposing it into irreducible stars:

$$\{k_1\} \times \{k_2\} = \sum_\delta n_\delta \{k_\delta\} \tag{3.16.8}$$

where initial wavevector of each star $\{k_\delta\}$ is defined by the relation:

$$k_1 + d_\delta k_2 = k_\delta + b \tag{3.16.9}$$

where b is a reciprocal lattice vector. Since each one of the vectors k_δ and prong of their stars may be also obtained by other vectors of stars $\{k_1\}$and $\{k_2\}$ the number of appearance of each star is greater than unity.

The group K_δ of the wavevector k_δ is larger than intersection group M_δ and may be decomposed into left cosets to it:

$$K_\delta = \sum_{\nu=1}^{n_\delta} p_\nu M_\delta, \; n_\delta = \frac{|K_\delta|}{|M_\delta|} \tag{3.16.10}$$

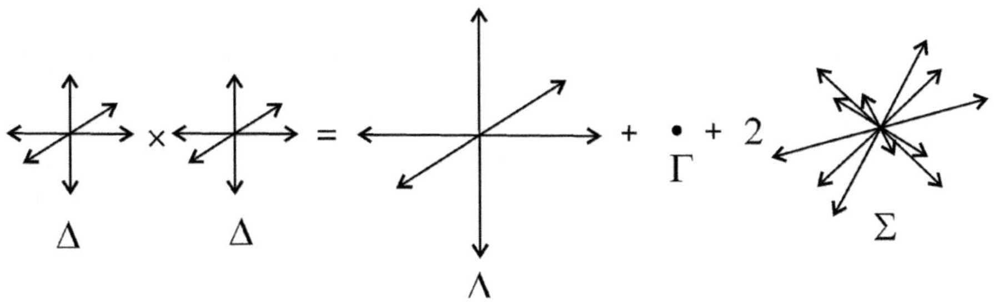

Fig. 3.16 Decomposition of the product of two wavevector stars $\{\Delta\} \times \{\Delta\}$ BZ is shown in Fig. 3.9

Using the transitivity of induction, we write Mackey theorem in the form:

$$\left(D_{(H)} \uparrow G\right) \times \left(B_{(F)} \uparrow G\right) = \sum_{\delta} \left(P^{\delta}_{(M_\delta)} \uparrow K_\delta\right) \uparrow G \tag{3.16.11}$$

Let us denote by C^i_δ the IRs of the group K_δ. Then the corresponding IRs of G are induced from K_δ and are written as $C^i_\delta \uparrow G$. Using Frobenius reciprocity theorem to decompose representations $P^{\delta}_{(M_\delta)} \uparrow K_\delta$ induced in K_δ, we obtain the final formula for the frequency of appearance of IR $C^i_\delta \uparrow G$ in the decomposition of Kronecker product:

$$f(C^i_\delta \uparrow G / \left(D_{(H)} \uparrow G\right) \times \left(B_{(F)} \uparrow G\right) = \frac{1}{|M_\delta|} \sum_{m \in M_\delta} \chi\left[D(m)\right] \chi\left[B(d_\delta^{-1} m d_\delta)\right] \chi^*\left[C^i_\delta(m)\right] \tag{3.16.12}$$

Formula (3.16.12) should be applied for each one of the double cosets δ in (3.16.4).

Example 3.15 Kronecker square of IRs $\Delta_i \uparrow G$ in direction Δ in a BZ of symmorphic space group O_h^5 (see Fig. 3.9).

Wavevector group H in direction Δ is a group $C_{4v} \times T$ (see Fig. 3.9). Since the translation subgroup is the same in all subgroups considered, it will not be mentioned below. We consider Kronecker square of one-dimensional IRs of C_{4v} induced into G. Direct product of two IRs in Δ−direction is envisaged in Fig. 3.16 by a direct product of stars. Decomposition of direct product of two stars $\{\Delta\}$ results in one star $\{\Delta\}$, six stars $\{\Gamma\}$ and two stars $\{\Sigma\}$. Star Σ consists of 12 prongs directed to midpoints of the edges of the cube. From the following consideration, it will be clear that the number of appearance of any star in the decomposition of direct product equals to the number n_δ in formula (3.16.10). To obtain the decomposition of Kronecker product, the group O_h should be decomposed into double cosets (3.16.4) with respect to two subgroups H. This decomposition consists of three double cosets, defined by identity element, space inversion and rotation C_3 about axis (111). For the double coset defined by the identity element IR P^δ (3.16.6) is a square of one-dimensional IR and equals to IR A_1 of C_{4v}. In notations for the space groups, we write Δ_1 instead of A_1. Thus the first term is written as $\Delta_1 \uparrow G$. For double coset defined by the space inversion we also obtain IR A_1 of C_{4v}, but in this case wavevector group in Γ point is a group O_h. For the Γ point making use of Frobenius theorem, we obtain that $A_{1(C_{4v})} \uparrow O_h = A_{1g} + E_g + T_{1u}$. For the star Σ we consider a prong $(0, k, k)$, which is the sum of two prongs $(0, 0, k)$ and $(0, k, 0)$ of the stars $\{\Delta\}$. The intersection of groups C_{4v} and $C_3 C_{4v} C_3^{-1}$ consists of σ_x and E. In our case D in (3.16.12) is IR of group C_{4v} with rotations about z-axis and $B(d_\delta^{-1} m d_\delta)$ is IR of C_{4v} with rotations about y-axis. For the element σ_x we must take character of $B\left(C_3^{-1} \sigma_x C_3\right) = B(\sigma_z)$. Since the reflections in both symmetry planes of C_{4v} belong to one class, the characters of $B(\sigma_z)$ and $D(\sigma_x)$ are the same. Thus for all one-dimensional IRs of C_{4v} in the initial state $\Delta_i \uparrow G$ we obtain totally symmetric IR A' of group $M_\delta = C_s$. Since the group C_{2v} of vector k_{Σ} is larger than M_{δ}, we must first induce into the former group: $A'_{(C_s)} \uparrow C_{2v} = A_1 + B_2$. Further induction into the space group results in its IRs. Thus the decomposition of Kronecker square is written as

$$\Delta_i \uparrow G \times \Delta_i \uparrow G = \Delta_1 \uparrow G + A_{1g} + E_g + T_{1u} + \Sigma_1 \uparrow G + \Sigma_4 \uparrow G$$

where Δ_i are one-dimensional IRs of C_{4v}, Δ_1 is IR A_1 of C_{4v}, Σ_1 and Σ_4 are IRs A_1 and B_2 of C_{2v}, respectively, and other notations correspond to point group O_h.

Chapter 4
Cooper Pairs as Two-Electron States in Crystals

I do not believe that the subject of the reduction of direct products of representations has been exhausted.

Wigner E.P. [64]

Abstract The chapter contains modern group-theoretical approach to the wavefunction of a Cooper pair, developed in recent years. Two-electron states in solids obeying Pauli exclusion principle are considered making use of Mackey theorem on symmetrized squares of induced representations. Such a state with zero total momentum corresponds to Cooper pairs in crystal solids. This approach, also called a space-group approach to the wavefunction of a Cooper pair, is applied to the groups D_{2h}, D_{4h} and D_{6h}, and T_h, i.e. symmetry groups of topological superconductors UTe_2, Sr_2RuO_4, UPt_3 and $PrOs_4Sb_{12}$ and possible pairs' symmetries in symmetry points, directions and planes in BZs are considered. Special attention is given for non-symmorphic groups, coupling with nonzero angular momentum and time-reversal symmetry. Additional quantum numbers for Cooper pairs belonging to two-dimensional IRs are introduced. The group theoretical approach is further developed to incorporate phase winding and the applications are considered.

4.1 Introduction

According to the BCS theory [65], two electrons in a Cooper pair have opposite spins and wavevectors k, and the pair creation operator is written as

$$b^{+} = c^{+}_{-k\downarrow} c^{+}_{k\uparrow} \tag{4.1.1}$$

where an arrow denotes a spin direction. Cooper pairs in BCS theory are singlet, completely symmetric in k-space, and invariant under the action of lattice translations, space inversion I and time-reversal ϑ. However, many of these features are violated in unconventional superconductors [66], for example, in UPt_3 [67–69], Sr_2RuO_4

V. G. Yarzhemsky, *Novel Group Theoretical Methods for Electron Structure Theory*,
https://doi.org/10.1007/978-3-031-72802-0_4

[70–73] and UTe_2 [74, 75], which are also called topological superconductors [76, 77]. The power-law temperature dependences of physical quantities (instead of exponential for BCS superconductors) indicate that the SOP (superconducting order parameter) vanishes at the points and on the lines on the FS (Fermi surfaces). The absence of changes in the Knight shift in the NMR spectra upon passing through T_c corresponds to the odd SOP, and hence to the triplet type of pairing. Muon spin relaxation experiments indicate the presence of spontaneous magnetic fields. Therefore, in many models, superconductivity in unconventional superconductors was explained by magnetic (spin) fluctuations. The recently discovered superconductor UTe_2 [74] is characterized not only by an extremely high critical field $H_{c2} = 30$ T, but also by the appearance of a new superconducting phase at $H = 60$ T supported by a magnetic field [75]. The experimental data discussed above point to the need of development of new theoretical models and symmetry approaches to describe the SOP in unconventional superconductors. A generalization of the symmetry principles for triplet pairing was proposed by Anderson [78]. It is assumed that the crystal has symmetry with respect to the inversion I and time reversal ϑ, thus two states with opposite spins and directions of the wavevector k are degenerate. Inversion changes the direction of the vector k, and time reversal changes the direction of the vector k and the spin projection. Thus, there is a fourfold degeneracy of states at a general point in a BZ (Brillouin zone):

$$\varphi_{k\uparrow} \tag{4.1.2}$$

$$I\varphi_{k\uparrow} = \varphi_{-k\uparrow} \tag{4.1.3}$$

$$\vartheta\varphi_{k\uparrow} = \varphi_{-k\downarrow} \tag{4.1.4}$$

$$\vartheta I\varphi_{k\uparrow} = \varphi_{k\downarrow} \tag{4.1.5}$$

Singlet pair state is a linear combination of two states (φ_k,$\vartheta\varphi_k$) and ($I\varphi_k$, $I\vartheta\varphi_k$), and pair's function is even with respect to space inversion. Pair functions (φ_k, $I\varphi_k$) and ($\vartheta\varphi_k$, $I\vartheta\varphi_k$) correspond to triplet states $M_s = 1, -1$, and the $M_s = 0$ component is an odd combination of (φ_k, $\vartheta\varphi_k$) and ($I\varphi_k$, $I\vartheta\varphi_k$).

In $L - S$ coupling scheme singlet and triplet pair functions are written as

$$\psi_k^s = (\varphi_k(r_1)\varphi_{-k}(r_2) + \varphi_k(r_2)\varphi_{-k}(r_1))\, S^0 \tag{4.1.6}$$

$$\psi_k^t = (\varphi_k(r_1)\varphi_{-k}(r_2) - \varphi_k(r_2)\varphi_{-k}(r_1)\, S_M^1, \quad M = -1, 0, 1 \tag{4.1.7}$$

where S^0 and S_M^1 denote the singlet and triplet spin functions. It can be seen from these formulas that in the singlet (triplet) state the spatial part of the pair wavefunction is even (odd) with respect to the permutation of coordinates r_1 and r_2. Since in this case the permutation of coordinates is equivalent to the action of the inversion on the function of the pair, the wavefunction of a singlet pair is even with respect to the inversion and the wavefunction of a triplet pair is odd. However, as will be shown in the next section, this relation is violated on the axes of symmetry of the third, fourth and sixth orders in k-space.

In this section, we consider the phenomenological representation [79–81, 83] of a Cooper pair wavefunction. In this approach, the wavevector k is considered in isotropic space. In k-space, spherical functions are used to represent pair's function. For example, pair's functions with angular momentum $l = 1$ are written as

$$k_z = |k|\, Y_0^1 = |k| \sqrt{\frac{3}{4\pi}} \cos\theta \tag{4.1.8}$$

$$k_x \pm i k_y = |k|\, Y_{\pm 1}^1 = |k| \sqrt{\frac{3}{8\pi}} \sin\theta e^{\pm i\varphi} \tag{4.1.9}$$

Also spherical functions of k with $l = 2$ and $l = 3$ and $l = 4$, similar to atomic $d-$, $f-$ and $g-$functions, are used. It was shown above that at the general point of the BZ singlet function of a pair is even and triplet function of a pair is odd. Spherical functions with even value of angular momentum are even with respect to inversion, and spherical functions with odd value of angular momentum are odd. Relying on these properties of the spherical function it is usually assumed in phenomenological approaches that singlet superconductors are described by $s-$, $d-$ and $g-$spherical functions, and triplet ones are described by $p-$ and $f-$functions [76]. However, according to group theory, every irreducible representation $D^{(l)}$ of a rotation group $\mathsf{O}(3)$, characterized by the quantum number l, has two extensions onto rotation group completed by the left coset $I\mathsf{O}(3)$: even $D^{(l)+}$ and odd $D^{(l)-}$. Therefore, the direct relation between the angular momentum and parity of a pair is not a fundamental physical relation, but is a consequence of the choice of spherical functions as a basis set for Cooper pairs.

In the triplet case, the SOP is usually expressed in terms of the so-called d-vector, which is a function of the momentum and the spin of an electron pair:

$$\Delta_{\sigma\sigma'}(\boldsymbol{k}) = (\boldsymbol{d}(\boldsymbol{k}) \cdot \boldsymbol{\sigma}) \begin{pmatrix} 0 & 1 \\ -1 & 0 \end{pmatrix} \tag{4.1.10}$$

where point is used for the "scalar" product of a $d-$vector and a "vector" constructed from Pauli matrices. This expression may be written also as

$$\Delta_{\sigma\sigma'}(\boldsymbol{k}) = \begin{pmatrix} -d_x(\boldsymbol{k}) + i d_y(\boldsymbol{k}) & d_z(\boldsymbol{k}) \\ -d_z(\boldsymbol{k}) & d_x(\boldsymbol{k}) + i d_y(\boldsymbol{k}) \end{pmatrix} \tag{4.1.11}$$

Thus SOP of triplet pair is written as a complex vector, whose direction is normal to the triplet pair spin direction. A state is called non-unitary if:

$$\left[\boldsymbol{d}(\boldsymbol{k}) \times \boldsymbol{d}^*(\boldsymbol{k})\right] \neq 0 \tag{4.1.12}$$

Using the SOP expression in terms of the $d-$vector, one obtains an equivalent definition of non-unitarity:

$$\left|\Delta_{\uparrow\uparrow}(\boldsymbol{k})\right|^2 - \left|\Delta_{\downarrow\downarrow}(\boldsymbol{k})\right|^2 = 2i\left[\boldsymbol{d}(\boldsymbol{k}) \times \boldsymbol{d}^*(\boldsymbol{k})\right]_z \neq 0 \tag{4.1.13}$$

The pair angular momentum is defined as

$$L = -i\left\langle \Psi * \left[\boldsymbol{k} \times \frac{\partial}{\partial \boldsymbol{k}}\right] \Psi \right\rangle \tag{4.1.14}$$

In spectroscopic notations, S_1^1 and S_{-1}^1 correspond to $\uparrow\uparrow$ and $\downarrow\downarrow$, and also S_0^1 and S^0 correspond to the symmetrized and antisymmetrized linear combinations of $\uparrow\downarrow$and $\downarrow\uparrow$. The triplet spin components are also written as spin vector components: $\hat{x} = \frac{1}{2}\left(S_1^1 + S_{-1}^1\right)$, $\hat{y} = \frac{1}{2i}\left(S_1^1 + S_{-1}^1\right)$ and $\hat{z} = S_0^1$. In these notations, the phases of superfluid ^{3}He are written as

$$\boldsymbol{d}(\boldsymbol{k}) = \hat{z}(k_x \pm ik_y) \text{ (A phase)} \tag{4.1.15}$$

$$\boldsymbol{d}(\boldsymbol{k}) = \hat{x}k_x + \hat{y}k_y + \hat{z}k_z \text{ (B phase)} \tag{4.1.16}$$

$$\boldsymbol{d}(\boldsymbol{k}) = \left(\hat{x} + i\hat{y}\right)\left(k_x + ik_y\right) \text{(A1 phase)} \tag{4.1.17}$$

Phases A and B are unitary, but the phase A1 is non-unitary.

The examples of phenomenological SOP for D_{4h} group (symmetry group of Sr_2RuO_4) are presented in Table 4.1. It should be noted that Sr_2RuO_4 initially considered as triplet superconductors [70, 71] now is reconsidered as singlet one [72] and there is a transition between unitary and non-unitary phases [72]. Many of the functions in Table 4.1 are considered as potential candidates for SOP of Sr_2RuO_4. However chiral functions of E_g and E_u symmetry in Table 4.1 have no nodes in vertical planes of chiral phase, but vertical nodal planes were manifested experimentally in Sr_2RuO_4 [84].

The basis functions for the two phases of UPt_3 are well represented by linear combinations of two spherical functions and transitions are described by adjusting

Table 4.1 Phenomenological pair functions for D_{4h} symmetry [79, 80]

	Singlet pair		Triplet pair
IR	Basis function	IR	Basis function
A_{1g}	$k_x^2 + k_y^2$	A_{1u}	$\hat{x}k_x + \hat{y}k_y$
A_{1g}	k_z^2	A_{1u}	$\hat{z}k_z$
A_{2g}	$k_x k_y \left(k_x^2 - k_y^2\right)$	A_{2u}	$\hat{x}k_y - \hat{y}k_x$
B_{1g}	$k_x^2 - k_y^2$	B_{1u}	$\hat{x}k_x - \hat{y}k_y$
B_{2g}	$k_x k_y$	B_{2u}	$\hat{x}k_y + \hat{y}k_x$
E_g	$k_z\left(k_x, k_y\right)$	E_u	$\hat{z}\left(k_x, k_y\right), k_z\left(\hat{x}, \hat{y}\right), \hat{z}(k_x \pm ik_y)$

Table 4.2 Triplet basis functions of Cooper pairs for D_{2h} point group [85]

IR	Basis functions
A_u	$k_x\hat{x}, k_y\hat{y}, k_z\hat{z}$
B_{1u}	$k_y\hat{x}, k_x\hat{y}, k_xk_yk_z\hat{z}$
B_{2u}	$k_x\hat{z}, k_z\hat{x}, k_xk_yk_z\hat{y}$
B_{3u}	$k_z\hat{y}, k_y\hat{z}, k_xk_yk_z\hat{x}$

of parameters η_1 and η_2 [68, 69]:

$$\boldsymbol{d}(\boldsymbol{k}) = \hat{z}k_z\left[\left(\eta_1(k_x^2 - k_y^2\right) \pm 2i\eta_2k_xk_y\right] \tag{4.1.18}$$

Since IRs of symmetry group D_{2h} of UTe_2 are one-dimensional and real, its phenomenological basis functions [85] are real and correspond to unitary SOP (see Table 4.2). To describe non-unitary SOP, complex linear combinations of different IRs were used (see Table 4.3). However, these functions being complex linear combinations of functions belonging to different IRs are not group-theoretical.

Thus phenomenological approaches qualitatively represent experimental data but general group-theoretical consideration is required to identify underlying symmetries and Cooper pairs' structures. According to Anderson ansatz (4.1.6), the wavefunction of a Cooper pair in $(k, -k)$ manifold is constructed as a function of two equivalent fermions, i.e. the spatial part of the wavefunction of a singlet (triplet) pair belongs to symmetrized (antisymmetrized) Kronecker of one-electron wavefunction (IR). Bradley and Davies [38] were the first who made use of Mackey theorem [36] and solved the problem of calculations of symmetrized (antisymmetrized) Kronecker squares of the space-group IRs and opened the way for exact quantum mechanical approaches to many-body effects in condensed matter physics.

The space-group approach to the wavefunction of a Cooper pair [40] is based on Anderson ansatz (4.1.6) and Mackey-Bradley-Davies technique [38]. This approach was further developed and applied to unconventional superconductors [40–45, 47, 86–94]. It was shown that on symmetry lines inside a BZ possible pair symmetry depends on the small IR of one-electron state and that a direct relation between multiplicity and parity of a pair is violated for two-dimensional small IRs on lines of symmetry. Also the results for non-symmorphic space groups differ from the results for symmorphic group on the surface of a BZ. A generalized approach has recently

Table 4.3 Triplet basis functions with angular momentum for UTe_2 in magnetic field [85]. In the presence of magnetic field along one of the axis, the SOP is expressed as complex linear combination of basis functions of different IRs

H	SOP
H_x	$A_u + iB_{3u}$, $B_{1u} + iB_{2u}$
H_y	$A_u + iB_{2u}$, $B_{1u} + iB_{3u}$
H_z	$A_u + iB_{1u}$, $B_{2u} + iB_{3u}$

been developed, which makes it possible to combine the Ginzburg-Landau topological phase winding with the symmetry of the space group [50, 51] and additional quantum numbers for a pair in axial symmetry group were introduced. It was also proposed to use magnetic point groups for construction of pair functions [49, 51].

In this chapter, the space-group approach to the wavefunction of a Cooper pair with its extension on topological phase winding is presented with numerical examples and illustrations for the symmetry groups of D_{6h}, D_{4h}, D_{2h} and T_h, i.e. symmetry groups of topological superconductors UPt_3, Sr_2RuO_4 and UTe_2 and $PrOs_4Sb_{12}$. Special attention is paid to non-symmorphic groups [19, 60] magnetic symmetry as well as electron pairs with nonzero angular momentum.

4.2 Space-Group Approach to the Wavefunction of a Cooper Pair

The space-group approach to the wavefunction of a Cooper pair is the generalization of Anderson ansatz (4.1.6) for a Cooper on real space-group symmetry. This approach makes it possible to take into account point group symmetry, magnetic group symmetry and improper translations of non-symmorphic space groups.

One-electron states in a crystal with a symmetry group G are labeled by the wavevector k, its symmetry group H (little group) and the index κ of small IR $D^{\kappa k}$ of H. In what follows we will assume that one index κ includes k, drop the index k and write D^{κ}. Consider a left coset decomposition of a space group with respect to H:

$$G = \sum_{\sigma} s_{\sigma} H \tag{4.2.1}$$

The acting of the left coset representatives s_{σ}on k results in all prongs of k_{σ} of the wavevector star $\{k\}$:

$$k_{\sigma} = s_{\sigma} k + b_{\sigma} \tag{4.2.2}$$

where b_{σ} is a reciprocal lattice vector. Thus in crystal solids one-electron states are defined by a wavevector star instead of two vectors k and $-k$ in the case of spherically symmetric Fermi liquid. The IR of the space group is an induced representation, defined by the following formula:

$$D^{\kappa} \uparrow G\ (g)_{\sigma i, \tau j} = D^{\kappa}_{i,j}(s_{\sigma}^{-1} g s_{\tau})\delta\left(s_{\sigma}^{-1} g s_{\tau}, H\right) \tag{4.2.3}$$

where σ and τ correspond to the left coset decomposition of the space group with respect to H (4.2.1) and i and j denote rows and columns of the matrix D^{κ}. In a general point of a BZ, the dimension of IR $D^{\kappa} \uparrow G$ equals to the number $\left|\bar{G}\right|$ of point group elements in $\bar{G}$ (central extension of G). Two-electron space is a Kronecker square of this space and its dimension in a general $k-$point equals to $|\bar{G}|^2$. This space can be easily decomposed into physically different parts making use of double coset decomposition of G with respect to H:

$$G = \sum_{\delta} H d_{\delta} H \tag{4.2.4}$$

The double coset representatives d_{δ} denote different terms in a Kronecker square $D^{\kappa} \uparrow G \times D^{\kappa} \uparrow G$ decomposition. The notation $\times$ for direct (Kronecker) product is used throughout. For each double coset δ a representation $P_{\delta}^{\kappa\pm}$ is considered, whose character is written as

$$\chi\left[P_{\delta}^{\kappa\pm}(m)\right] = \chi\left[D^{\kappa}(d_{\delta}^{-1} m d_{\delta})\right] \chi\left[D^{\varkappa}(m)\right] \tag{4.2.5}$$

where m are the elements of the intersection group M_{δ}, defined by the relation:

$$M_{\delta} = H \cap d_{\delta} H d_{\delta}^{-1} \tag{4.2.6}$$

Consider the case of nontrivial self-inverse double coset, that is, if:

$$H d_{\alpha} H = H d_{\alpha}^{-1} H \neq H \tag{4.2.7}$$

There must exist elements h_1, h_2, h_3 and h_4 such that $h_1 d_{\alpha} h_2 = h_3 d_{\alpha}^{-1} h_4$, i.e. $d_{\alpha} h_2 h_4^{-1} = h_1^{-1} h_3 d_{\alpha}^{-1}$. Thus the intersection $d_{\alpha} H \cap H d_{\alpha}^{-1}$ is nonempty and we can choose an element a from this intersection:

$$a = d_{\alpha} h = \bar{h} d_{\alpha}^{-1}, \; h, \; \bar{h} \subset H \tag{4.2.8}$$

Thus the extended intersection group is introduced:

$$\tilde{M}_{\alpha} = M_{\alpha} + a M_{\alpha} \tag{4.2.9}$$

and then M_{α} is an invariant subgroup of index 2 of $\tilde{M}_{\alpha}$.

There are two extensions of $P_{\alpha}^{\kappa\pm}$ into group $\tilde{M}_{\alpha}$, corresponding to symmetrized and antisymmetrized squares , respectively:

$$\chi\left[P_{\alpha}^{\kappa+}(am)\right] = +\chi\left[D^{\kappa}(amam)\right], \tag{4.2.10}$$

$$\chi\left[P_{\alpha}^{\kappa-}(am)\right] = -\chi\left[D^{\kappa}(amam)\right] \tag{4.2.11}$$

According to Mackey-Bradley theorem, symmetrized (square brackets) and antisymmetrized (curly brackets) parts of the Kronecker square of induced representation are written as

$$\left[D^{\kappa}\uparrow G\times D^{\kappa}\uparrow G\right]=\left[D^{\kappa}\times D^{\kappa}\right]\uparrow G+\sum_{\alpha}P_{\alpha}^{\kappa+}\uparrow G+\sum_{\beta}P_{\beta}^{\kappa}\uparrow G \quad (4.2.12)$$

$$\{D^{\kappa}\uparrow G\times D^{\kappa}\uparrow G\}=\{D^{\kappa}\times D^{\kappa}\}\uparrow G+\sum_{\alpha}P_{\alpha}^{\kappa-}\uparrow G+\sum_{\beta}P_{\beta}^{\kappa}\uparrow G \quad (4.2.13)$$

The symmetrization (antisymmetrization) in the first item is performed by standard point group technique, considered in Chap. 1, on subgroup H, sum in the second item runs over all self-inverse double cosets and the sum in the third item runs over all not self-inverse double cosets, i.e. if $Hd_{\beta}H\neq Hd_{\beta}^{-1}H$.

Total momentum of resulting electron pair depends on double coset representative and is written as

$$K_{\alpha}=\ k+d_{\alpha}k+b_{\alpha} \quad (4.2.14)$$

where b_{α} is a reciprocal lattice vector. In a general case, the group L_{α} of the vector K_{α} is larger than the group $\tilde{M}_{\alpha}$. Making use of transitivity of induction theorem, we induce the representations $P_{\alpha}^{\kappa\pm}$ firstly into L_{α}. The induced representations $P_{\alpha}^{\kappa\pm}\uparrow L_{\alpha}$ can be easily decomposed into irreducible representations B_{α}^{j} of group L_{α} making use of Frobenius reciprocity theorem:

$$f_{\alpha\kappa\pm}^{j}=\frac{1}{\left|\tilde{M}\right|}\sum_{\tilde{m}\in\tilde{M}}\chi^{*}\left(B_{\alpha}^{j}\left(\tilde{m}_{\alpha}\right)\right)\chi\left(P_{\alpha}^{\kappa\pm}\left(\tilde{m}_{\alpha}\right)\right) \quad (4.2.15)$$

Further induction of IRs into the whole space group G results in irreducible representations $B_{\alpha}^{j}\uparrow G$.

In the case of non-symmorphic space-group formulas (4.2.5–4.2.11) should be written with account for improper translation:

$$\chi\left[P_{\alpha}^{\kappa\pm}\ \{m|\tau_{m}\}\right]=\chi\left[D^{\kappa}(\{d|\tau_{d}\}^{-1}\{m|\tau_{m}\}\{d|\tau_{d}\}\right]\chi\left[D^{\kappa}\left(\{m|\tau_{m}\}\right)\right], \quad (4.2.16)$$

$$\chi\left[P_{d}^{\kappa\pm}\left(\{a|\tau_{a}\}\ \{m|\tau_{m}\}\right)\right]=\pm\chi\left[D^{\kappa}(\{a|\tau_{a}\}\ \{m|\tau_{m}\}\{a|\tau_{a}\}\ \{m|\tau_{m}\})\right] \quad (4.2.17)$$

where subscript α of d is dropped. The IRs D^{κ} of the wavevector group H are connected with the projective representations of its central extension H as: $\hat{D}^{k}(h)=\exp(ik\cdot(\tau_{h}))\,D^{\kappa}\,\{h|\tau_{h}\}$. Its factor system is defined by the elements of H. It should be noted that since in formula (4.2.17) some elements of H multiplied by an element $\{a|\tau_{a}\}\notin H$, we can't use projective representations in (4.2.17). In this case, the IRs D^{κ} wavevector group H should be used in calculations.

In the case of Cooper pairs, the double coset representative d_{α} is a space inversion I and the $a=I$ in (4.2.8). Also in some symmetrical points on the surface of a BZ d_{α} is an identity element E. In this case, zero total momentum of a pair is achieved due to translation periodicity.

According to Pauli exclusion principle, symmetrized square $\left[\hat{D}^{\kappa} \uparrow G \times \hat{D}^{\kappa} \uparrow G\right]$ of spatial part of the wavefunction corresponds to a singlet pair and antisymmetrized square corresponds to a triplet pair. In the case of strong spin-orbit coupling, total pair's wavefunction belongs to antisymmetrized square of double-valued IR $\left\{D'^{\kappa} \uparrow G \times D'^{\kappa} \uparrow G\right\}$.

In the case of Cooper pairs $K_\alpha = 0$ in formula (4.2.14) and the induced representation $P_\alpha^{\kappa\pm} \uparrow \hat{G}$ is a reducible representation of a point group $\bar{G}$, which is a central extension of G and in B_α^j formula (4.2.15) is IR of point group $\bar{G}$.

Mackey theorem makes it possible to draw a number of conclusions regarding the possible symmetries of Cooper pairs. For k at a general point in the BZ, the group M_α consists of the identity element only, and the group $\tilde{M}_\alpha = \{E, I\}$. The characters of the representations are $P_\alpha^{\kappa+}$ and $P_\alpha^{\kappa-}$ for α being a space inversion I equal to $+1$ and -1, respectively, and for the identity element both characters equal to $+1$ (see Table 4.4). Using the Frobenius reciprocity theorem, we obtain that in a general point of the BZ (one electron), all even IRs are possible for singlet pairs, and all odd IRs are possible for spatial parts of triplet pairs. Moreover, each even (odd) IR of the point group enters the expansion of symmetrized (antisymmetrized) square as many times as its dimension. Thus, it also follows from Mackey theorem that in a general point of the BZ the dimensions of the spaces of singlet pairs and of spatial parts of triplet pairs coincide and are equal to $\bar{G}/2$. In the $L - S$ coupling scheme, the spatial part of a triplet pair is multiplied by the three components of triplet spin $\hat{x}$, $\hat{y}$ and $\hat{z}$ and the dimension of the space of total triplet pairs' wavefunctions for k a general point in a BZ is equal to $3\bar{G}/2$.

In a strong spin-orbit coupling case, the spin is included in one-electron wavefunction and the pair wavefunction is calculated as antisymmetrized square of double-valued IR of the space group. However in k-points of low-symmetry additional degeneracy due to time-reversal symmetry should be taken into account. The magnetic little co-group consists of the elements E and I. The sum in Herring criterion consists of one element $(\vartheta I)^2$ and since $\nu(\vartheta^2) = -1$, this sum equals to -1 and the corepresentation belongs to the case (b). Thus the character of corepresentation equals to 2. The character of antisymmetrized square of double-valued IR with account for time reversal $P_I'^{-}$ on group C_i is presented in Table 4.4. Spin functions $\hat{x}$, $\hat{y}$ and $\hat{z}$ belong to totally symmetric IR of C_i. Hence, it follows from Table 4.4 that sum of IRs of singlet and triplet pairs with account for spin function equals to the sum of possible IRs in the case of strong spin-orbit coupling case.

Inside the BZ in the majority of cases (but not in all) spatial part of a singlet pair is even and spatial part of a triplet pair is odd. If IR D^κ is two-dimensional (it takes place on the threefold, fourfold and sixfold axes), one obtains from (4.2.5), (4.2.10), (4.2.11) that $\chi\left(P_I^{\kappa\pm}(E)\right) = 4$, $\chi\left(P_I^{\kappa+}(I)\right) = 2$ and $\chi\left(P_I^{\kappa-}(I)\right) = -2$. Hence it follows that in this case even and odd IRs are mixed in the decomposition of symmetrized and antisymmetrized squares.

In axial symmetry groups D_{2h}, D_{4h} and D_{6h} the little group H coincides with the whole group on the surface of a BZ at point $k_z = b_3/2$ and there in one double coset only defined by the identity element. In this case, two electrons with the same

Table 4.4 Characters of possible pairs' functions on group C_i (for k a general point in a BZ). P_I^+ and P_I^- are spatial parts of singlet and triplet pairs and P'^- is total wavefunction, constructed on the basis of double-valued IR with account for time-reversal symmetry

P^j	E	I	IR
P_I^+	1	1	A_g
P_I^-	1	-1	A_u
$P_I'^-$	4	-2	$A_g + 3A_u$

momentums form a pair and zero total momentum is ensured by the translational periodicity. This case corresponds to the first item in Mackey theorem (4.2.12–4.2.13) and the characters of symmetrized and antisymmetrized squares are written by the standard way:

$$\chi\left[D^\kappa(g) \times D^\kappa(g)\right] = \frac{\chi^2\left(D^\kappa(g)\right) + \chi\left(D^\kappa\left(g^2\right)\right)}{2} \tag{4.2.18}$$

$$\chi\left\{D^\kappa(g) \times D^\kappa(g)\right\} = \frac{\chi^2\left(D^\kappa(g)\right) - \chi\left(D^\kappa\left(g^2\right)\right)}{2} \tag{4.2.19}$$

In this case, spatial parts of possible singlet and triplet pairs are even.

If the space group is non-symmorphic the symmetrized (antisymmetrized) characters for point $k_z = b_3/2$ should be written with account for factor system of projective representation $\hat{D}^\kappa$:

$$\chi\left[\hat{D}^\kappa(g) \times \hat{D}^\kappa(g)\right] = \frac{\chi^2\left[\hat{D}^\kappa(g)\right] + \omega(g, g)\chi\left[\hat{D}^\kappa\left(g^2\right)\right]}{2} \tag{4.2.20}$$

$$\chi\left\{\hat{D}^\kappa(g) \times \hat{D}^\kappa(g)\right\} = \frac{\chi^2\left[\hat{D}^\kappa(g)\right] - \omega(g, g)\chi\left[\hat{D}^\kappa\left(g^2\right)\right]}{2} \tag{4.2.21}$$

where $\omega(g, g)$ stands for the factor system of projective representation $\hat{D}^\kappa$. Due to the factors $\omega(g, g)$ even and odd IRs are mixed in this case.

Also it may be derived from formulas (4.2.12) and (4.2.13) that a usually accepted assertion that triplet (singlet) pairs have odd (even) angular momentum value is violated. For example, consider O_h group and k a general point in a BZ. In this case, all even (all odd) IRs appear in the decomposition for singlet (triplet) pairs with frequencies equal to their dimensions. Consider IR T_{1g} for a singlet pair and T_{2u} and E_u for triplet pairs. It is known that even spatial part of the wavefunction of atomic term $p^2(^3P)$ transforms as T_{1g} in O_h symmetry. Also odd spatial part of atomic term $p^3(^2D)$ splits in crystal field into T_{2u} and E_u. Hence, it follows that the angular momentum $l = 1$ may be assigned to a singlet pair of T_{1g} symmetry and angular momentum $l = 2$ may be assigned to triplet pairs of T_{2u} and E_u symmetry.

4.3 Pairs on Symmetry Planes in a BZ

Consider a pair on a symmetry plane beyond a symmetry axis inside a BZ (see Fig. 4.1). The wavevector group H consists of two elements E and σ_h, and group $\tilde{M}_\alpha$ is a group C_{2h} which includes also I and C_2. The characters of symmetrized and antisymmetrized squares together with the characters of group C_{2h} are presented in Table 4.5. It is seen from Table 4.5 that the symmetry of singlet pair is A_g and the symmetry of spatial part of triplet pair is B_u and that IRs B_g and A_u are prohibited on plane. It is clear from Frobenius reciprocity theorem that when IRs A_g (B_u) will be induced into G not all even (odd) IRs will be possible for singlet (triplet) pairs. However, total wavefunction of a triplet pair includes spin function, which may be represented as even vector ($\hat{x}$, $\hat{y}$, $\hat{z}$). The spin component normal to the plane (in

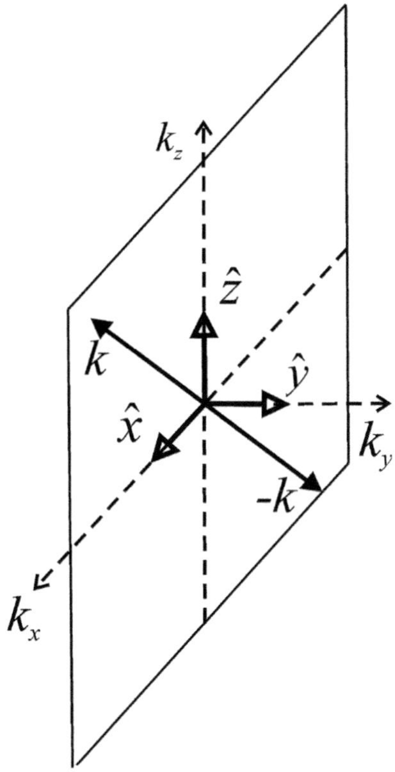

Fig. 4.1 Triplet two-electron state on a symmetry plane inside a BZ

Table 4.5 Characters of IRs of group C_{2h}, and of spatial part of singlet pair P^+, triplet pair P^- and triplet pair multiplied by spin part $(A_g + 2B_g)$ (see Fig. 4.1 and text)

IR	E	σ_h	I	C_2
A_g	1	1	1	1
B_g	1	-1	1	-1
A_u	1	-1	-1	1
B_u	1	1	-1	-1
$P^+ = A_g$	1	1	1	1
$P^- = B_u$	1	1	-1	-1
$P^- \times A_g = B_u$	1	1	-1	-1
$P^- \times 2B_g = 2A_u$	2	-2	-2	2

our case $\hat{y}$) transforms as A_g, and in plane spin components (in our case $\hat{x}$ and $\hat{z}$) transform as B_g. When multiplying spatial part of a triplet pair B_u by A_g and B_g we obtain B_u and A_u, respectively. Thus total wavefunctions of triplet pairs may include all odd IRs of C_{2h} (see Table 4.5). Consider the case when some IR Γ^q of the whole group corresponds to Cooper pairs at general k-point in a BZ, but is absent on the symmetry plane in a BZ. The intersection of the plane with the FS (Fermi surface) results in a line. Thus, a nodal plane in a BZ corresponds to a nodal line of SOP on FS. Therefore we obtained that for singlet pairs on planes some IRs of the whole group are prohibited and symmetry protected lines of nodes of SOP are possible. On the other hand, total wavefunctions of triplet pairs on plane include all odd IRs of C_{2h} and there are no symmetry protected lines of nodes (Blount theorem [95]).

The example of application of this technique to D_{4h} group is presented in Table 4.6. It can be seen from this table that when the spin of triplet pair is taken into account, there is no complete prohibition on pairs of any symmetry on planes. This result is in agreement with Blount theorem. However, in the case of axial symmetry, the spin functions $\hat{z}$ and $(\hat{x}, \hat{y})$ are physically different. OSP (opposite spin pairing) spin part $\hat{z}$ is transformed according to the IR A_{2g} of D_{4h}, and ESP (equal spin pairing) spin part $(\hat{x}, \hat{y})$ is transformed according to the IR E_g. Usually, for some particular triplet superconductor, on the basis of experimental data, a certain model

Table 4.6 Symmetry of total triplet pair functions on the planes of symmetry of group D_{4h}

	Spin	IR of $\tilde{M}_\sigma$	IRs of G
(001)	$\hat{z}$	B_u	$2E_u$
	$\hat{y}$	A_u	$A_{1u} + A_{2u} + B_{1u} + B_{2u}$
	$\hat{x}$	A_u	$A_{1u} + A_{2u} + B_{1u} + B_{2u}$
(010)	$\hat{z}$	A_u	$A_{1u} + B_{1u} + E_u$
	$\hat{y}$	B_u	$A_{2u} + B_{2u} + E_u$
	$\hat{x}$	A_u	$A_{1u} + B_{1u} + E_u$

is accepted: OSP or ESP. In this case, group-theoretical line of nodes for triplet SOP are possible. It is seen from Table 4.6 that in OSP model IR A_{2u} has zeros in all vertical planes. In basal plane, OSP pairs can belong to E_u only, and ESP pairs in basal plane may belong to any of one-dimensional IRs.

The emergence of nodal planes can be visualized using general form of the pair's function. Total momentum a Cooper pair equals to zero and singlet (triplet) wavefunction is symmetric (antisymmetric) with respect to permutations of electronic coordinates:

$$\psi^{s(t)} = \varphi_k(r_1)\varphi_{Ik}(r_2) \pm \varphi_{Ik}(r_1)\varphi_k(r_2) \tag{4.3.1}$$

where $Ik = -k$. In what follows, we assume that the coordinate number corresponds to the position of the one-electron wavefunction in the product and will omit it. In such a notation, the spatial part of a singlet (triplet) pair is written as

$$\psi^{s(t)} = \varphi_k\varphi_{Ik} \pm \varphi_{Ik}\varphi_k \tag{4.3.2}$$

The wavevector k is taken in a basis domain of a BZ, and the other basis functions are obtained by the action of the left coset representatives (4.2.1) on it. When projecting this basis set on IRs of point group the signs of the wavefunctions on both sides of the plane may be the same or different. The projection coefficients depend on the IR and define the signs in different sectors of a BZ. There are four distinct signs distributions on the plane, shown in Fig. 4.2. If the signs are the same, the functions of two pair's components on different faces merge on a plane and there are no nodes (see Fig. 4.2a, b for IRs A_g and B_u). Also if the signs on two sides of the plane are different two wavefunctions cancel on the plane, resulting in nodes (see Fig. 4.2c, d for B_g and A_u). Thus we obtained that if under the action of plane reflection σ_h and C_2 rotation about the axis normal to the plane spatial part of singlet (triplet) pair transforms according to A_g (B_u) of C_{2h}, the pair is nodeless in the plane. Also if under the action of plane reflection σ_h and C_2 rotation about the axis normal to the plane spatial part of singlet (triplet) pair transforms according to B_g (A_u) of C_{2h} then the pair function is nodal in this plane. Similar analysis can be performed for total triplet function, which includes spin.

Thus we have shown that in the case of real pairs functions the model of interference of pairs functions on two sides of symmetry plane gives the same results as Mackey-Bradley theorem. We will call such nodes group-theoretical nodes. However, it will be shown below that not all the nodes of two-dimensional IRs are directly connected with the label of IR. In this case, the nodes are defined by additional quantum numbers and may be analyzed by interference model considered in this section. In the subsequent sections, the interference model will be extended also on non-unitary (complex) pair functions.

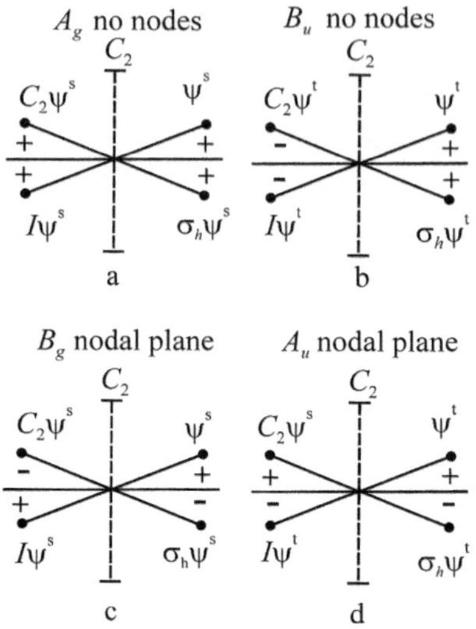

Fig. 4.2 Pair functions near symmetry plane (see text)

4.4 Faces of a BZ in Non-symmorphic Space Groups

In non-symmorphic space groups due to the presence of improper translations IRs on the surfaces of a BZ differ from those on the planes inside a BZ, resulting in different symmetry of two-electron states. Consider symmetry planes normal to vector k_z, passing through points $k_z = 0$ inside a BZ and $k_z = b_3/2$ on the face of a BZ, and take as an example symmorphic space group $D_{6h}^1(\,P6/mmm)$ and non-symmorphic space group $D_{6h}^4\,(\,P6_3/mmc)$ (see points P and Q in Fig. 4.25). At points P and Q the wavevector group consists of an identity element E and a reflection σ_z. The elements $\{\sigma_z|\tau_z\}$ and $\{C_{2z}|\tau_z\}$ of D_{6h}^4 are connected with improper translation $\tau_z = t_3/2$, where t_3 is the Bravais lattice translation. At the same time inversion $\{I|0\}$ appears without improper translation. It was shown in Sect. 3.13, that due to time-reversal symmetry double-valued ICRs of group D_{6h}^4 in basal plane and of group D_{6h}^1 in both horizontal planes are degenerated and belong to the type (c). Also it was obtained that in the plane $k_z = b_3/2$ in group D_{6h}^4 double-valued ICRs belong to type (b).

The extension of group G with time-reversal elements ϑG corresponds to paramagnetic state, in which both spin directions are equivalent. In this case, possible states are obtained as antisymmetrized squares of double-valued ICRs. Possible spin orientations of two sublattices connected by translation τ_z are shown in Fig. 4.3. In paramagnetic case, both spin directions are possible in two sublattices (see Fig. 4.3a). In ferromagnetic case, time reversal is forbidden and only one spin direction on two sublattices is possible (see Fig. 4.3b). In ferromagnetic case, possible two-electron states are obtained as antisymmetrized squares of double-valued IRs without taking into account additional degeneracy due to time-reversal symmetry.

Consider modifications of formulas for calculations of antisymmetrized Kronecker square (4.2.5–4.2.11) due to non-symmorphic structure of the group. Here and

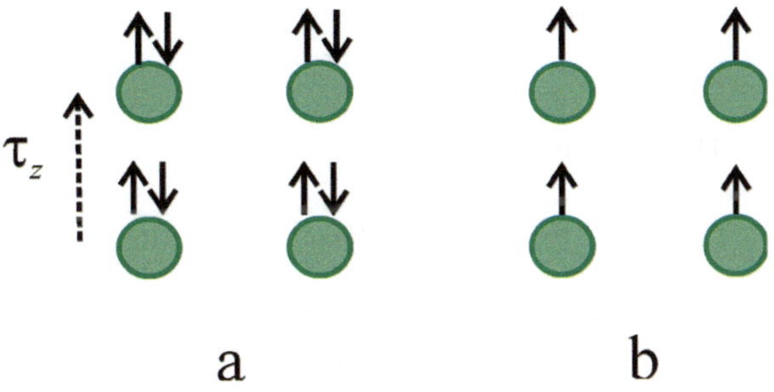

Fig. 4.3 Paramagnetic (**a**) and ferromagnetic (**b**) structures

throughout Kovalev's [60] notations $\hat{p}_i$ are used for double-valued projective IRs. Corresponding IRs of the space group are written as p_i. Corepresentations obtained from these IRs are denoted as $\Delta(p_i)$. Then the factor system ω_τ of projective representation is defined as follows:

$$\omega_\tau(\alpha, \beta) = \exp(ib \cdot \tau_\beta) \tag{4.4.1}$$

where the reciprocal lattice vector b is defined from the relation:

$$k - \alpha^{-1}k = b \tag{4.4.2}$$

It is clear from this formula that for $\alpha = \beta = \sigma_z$ the phase factor for the product σ_z^2 is not trivial, if improper translation τ_β is in $z-$direction. The factors ω can be obtained from the tables of projective IRs [60], since:

$$\hat{D}(\alpha)\hat{D}(\alpha) = \hat{D}(\alpha^2)\omega(\alpha, \alpha), \tag{4.4.3}$$

where the phase factor for the single-valued IRs $\omega(\alpha, \alpha) = \omega_\tau(\alpha, \alpha)$ is defined by formula (4.4.1).

In the case of double-valued IRs $\omega(\alpha, \alpha) = \omega_\tau(\alpha, \alpha)\,\omega_s(\alpha, \alpha)$, where the last factor corresponds to half-integer spin.

Consider the influence of improper translations on the phase factors of the characters of $P^{\kappa-}$ making use of the multiplication rule for the space-group elements:

$$\{\alpha|\tau_\alpha\}\left\{\beta|\tau_\beta\right\} = \left\{\alpha\beta|\tau_\alpha + \alpha\tau_\beta\right\} \tag{4.4.4}$$

In the case of symmetry plane, there are four terms of $P^{\kappa-}$(4.2.5), (4.2.11), which will be considered separately. For the identity element the result is obvious:

$$\chi\left[P^{\kappa-}(E)\right] = \chi^2\left[D^\kappa(E)\right] \tag{4.4.5}$$

For the term $\chi\left[P^{\kappa-}(\sigma_z)\right] = \chi\left[D^\kappa(\sigma_z)\right]\chi\left[D^\kappa(I\sigma_z I)\right]$ we can write the product of the space-group elements in the second multiplier as

$$\{I|\tau_I\}\{\sigma_z|\tau_\sigma\}\{I|\tau_I\} = \{I|\tau_I\}\{C_{2z}|\tau_\sigma + \sigma_z\tau_I\} = \{\sigma_z|\tau_I - \tau_\sigma - \sigma_z\tau_I\} = \{E|\tau_I - 2\tau_\sigma - \sigma_z\tau_I\}\{\sigma_z|\tau_\sigma\} \tag{4.4.6}$$

Thus we have

$$\chi\left[P^{\kappa-}\{\sigma_z|\tau_\sigma\}\right] = \exp(-ik\cdot(\tau_I - 2\tau_\sigma - \sigma_z\tau_I))\chi^2\left[D^\kappa\{\sigma_z|\tau_\sigma\}\right] \tag{4.4.7}$$

In our case $\tau_I = 0$, $\tau_\sigma = t_3/2$, and since $k = b_3/2$ the exponential factor results in $\omega_\tau = -1$. Note that this formula is written for the space-group IRs and we should convert projective representations $\hat{D}^k(\alpha)$ from Kovalev's tables into IRs $D^k\{\alpha|\tau_\alpha\}$ of the wavevector group:

$$D^k\{\alpha|\tau_\alpha\} = \hat{D}^k(\alpha)\exp(-ik\cdot\tau_\alpha) \tag{4.4.8}$$

Since in our case $k = b_3/2$ and $\tau_\alpha = t_3/2$, we obtain that $\exp(-ik\cdot\tau_\alpha) = -i$. Thus we obtain the additional factor $\omega_p = (-i)^2 = -1$, which appears in this term if projective IRs are used. Thus two factors cancel and the second term is written as

$$\chi\left[P^{\kappa-}\{\sigma_z|\tau_\sigma\}\right] = \chi^2\left[\hat{D}^k(\sigma_z)\right] \tag{4.4.9}$$

Note that Micklitz and Norman [42, 43] used ordinary IRs of the wavevector group and their result for this term is written as

$$\chi\left[P^{\kappa-}\{\sigma_z|\tau_\sigma\}\right] = -\chi^2\left[D^k(\sigma_z)\right] \tag{4.4.10}$$

For $k_z = 0$ in D_{6h}^4 symmetry and for symmorphic group D_{6h}^1 both factors vanish and in both approaches the factor before χ^2 equals to $+1$.

For the space inversion, we have

$$\chi\left[P^{\kappa-}\{I|\tau_I\}\right] = -\chi\left[D^\kappa(\{I|\tau_I\}\{I|\tau_I\})\right] = -\chi\left[D^\kappa\{E|\tau_I + I\tau_I\}\right] = -\chi\left[(D^\kappa(E)\right] \tag{4.4.11}$$

Here and throughout the minus sign, corresponding to antisymmetrized square is shown explicitly. It follows from this formula that in non-symmorphic case there are no additional phase factors for $P^{\kappa-}\{I|\tau_I\}$.

For the remaining term for $D(ImIm)$, one can write

$$\{I|\tau_I\}\{\sigma_z|\tau_\sigma\}\{I|\tau_I\}\{\sigma_z|\tau_\sigma\} = \{I|\tau_I\}\{\sigma_z|\tau_\sigma\}\{C_{2z}|\tau_I - \tau_\sigma\} = \{I|\tau_I\}\{I|\tau_\sigma + \sigma_z\tau_I - \sigma_z\tau_\sigma\} = \{E|\tau_I - \tau_\sigma - \sigma_z\tau_I + \sigma_z\tau_\sigma\} \tag{4.4.12}$$

It should be noted that when expanding the term $D(ImIm)$, one should take into account that in the case of double-valued IRs additional factor $\omega_s(Im, Im)$ appears. In our case, $\tau_I = 0$ and $\sigma_z\tau_\sigma = -\tau_\sigma$. Thus the formula for the remaining term is written as

$$\chi\left[P^{\kappa-}(C_{2z}\right] = -\exp(-ik_z \cdot (-2\tau_\sigma))\chi\left[D^{\kappa}(E)\right]\omega_s(C_{2z}, C_{2z}) \tag{4.4.13}$$

In our case, the exponential factor equals to -1. The last projective factor ω_s appears for double-valued IRs since the product C_{2z}, C_{2z} results in a rotation by 2π and equals to -1.

In the case of grey Shubnikov group, it is sufficient to perform symmetrization of corepresentations over unitary subgroup. Hence, it follows that the formulas (4.2.5), (4.2.11) and formulas of the present section are applicable to corepresentations. The characters of projective corepresentations are $\chi\left(\hat{\Delta}(C_{2z})\right) = \pm 2$ for group D_{6h}^4 at $k = b_3/2$ and $\chi\left(\hat{\Delta}(C_{2z})\right) = 0$ for D_{6h}^4 in basal plane and for D_{6h}^1 (See Table 3.13). It is clear that $\chi\left(\hat{\Delta}(E)\right) = 2$ in all cases. When collecting all phase factors and transforming projective IR for C_{2z} into IRs of the space group we obtain the results presented in Table 4.7.

The results of the present section make it possible to draw following conclusions one the structure of two-electron states on symmetry planes of BZ. Note that on symmetry plane wavevector group C_s includes two elements E and σ_z and its extension by I results in group C_{2h}, which has four IRs A_g, B_g, A_u and B_u. We consider the case when translation τ_z is connected with the elements σ_z and C_{2z}.

1. In symmorphic and non-symmorphic space groups on symmetry plane inside a BZ and in symmorphic groups on a zone faces, at $b_3/2$ the symmetry of singlet pairs is A_g and possible symmetries of triplet pairs are $A_u + 2B_u$. Since both odd IRs are possible there are no symmetry protected nodes of triplet pairs on the symmetry planes (Blount theorem [95]).

2. In non-symmorphic space groups on symmetry planes at zone face $b_3/2$ symmetry protected lines of nodes for triplet pairs are possible. In the considered case,

Table 4.7 Antisymmetrized representations P^{p-} and corepresentation $P^{\Delta(p)-}$ in planes $k_z = 0$ and $k_z = b_3/2$ for groups D_{6h}^1 and D_{6h}^4, constructed by Mackey theorem (4.2.5), (4.2.11). Additional phase factors ω_s, ω_τ and ω_p are also presented. In the first case, there is only one phase factor ω_s. In the second case, there is a product of all three factors. Double-valued ICRs $\Delta(\hat{p}_i)$ are constructed in Sect. 3.13 and their types are indicated in the parenthesis. Double-valued IRs $\hat{p}_{1,2}$ are also presented

		E	$\{\sigma_z\|\tau_z\}$	I	$\{C_{2z}\|\tau_z\}$	IRs of C_{2h}
$k_z = 0$	$\hat{p}_{1,2}$	1	$\pm i$			
D_{6h}^1 ($k_z = 0$ and $k_z = b_3/2$)	P^{p-}	1	-1	-1	1	A_u
and D_{6h}^4 ($k_z = 0$)	$P^{\Delta(p)-}(c)$	4	0	-2	2	$A_g + B_u + 2A_u$
	ω_s	1	1	1	-1	
D_{6h}^4 ($k_z = b_3/2$)	$\hat{p}_{1,2}$	1	± 1			
	P^{p-}	1	1	-1	-1	B_u
	$P^{\Delta(p)-}(b)$	4	4	-2	-2	$A_g + 3B_u$
	ω_τ	1	-1	1	1	
	ω_p	1	-1	1	-1	

Table 4.8 Multiplication table $h_i \times h_j = h_k$ for half of the elements of group D_{4h}. Here i is the number in a row in the table, j is the number in a column and k is written in the table. The asterisk at h_k^* denotes that $h_i \times h_j = h_k + C_{2\pi}$. In this case for half-integer spin $\hat{D}(h_i)\hat{D}(h_2) = -\hat{D}(h_k)$. The remainder of multiplication table may be obtained by formula (4.5.1) for multiplication by the inversion I

1	2	3	4	13	14	15	16
2	1*	4*	3	14	13*	16*	15
3	4	1*	2*	15*	16*	13	14
4	3*	2	1*	16*	15	14*	13
13	15*	14	16	1*	3*	2	4*
14	16	13*	15	2*	4	1*	3*
15	13	16	14*	3	1*	4*	2*
16	14*	15*	13*	4	2	3	1*

IRs $A_g + 3B_u$ are possible and IRs B_g and A_u are forbidden. It follows from Frobenius reciprocity theorem that when A_g and B_u are induced into $\hat{G}$ some even and some odd IRs will be forbidden.

4.5 Cooper Pairs in D_{4h} Symmetry

In the present section, we consider the method of finding possible symmetries of singlet and triplet two-electron states using the example of crystals with a point group D_{4h}. To find products of point group elements, it is convenient to use the multiplication table of elements given in Kovalev book [60] in digital notation. Standard and digital notations are shown in Fig. 4.4. In digital notation, the 48 elements of the cubic group are numbered from 1 to 48. In this notation, group's elements h_i for $i \leq 24$ are pure rotations and h_{25} is a space inversion I. Multiplication by the inversion h_{25} is given by two following formulas:

$$h_{25}h_i = \begin{cases} h_{i+24} & \text{if } 1 \leq i \leq 24 \\ h_{i-24} & \text{if } 25 \leq i \leq 48 \end{cases} \tag{4.5.1}$$

Part of multiplication table [60] for rotational elements only of D_{4h} group is presented in Table 4.8. Making use the above rules for multiplication by space inversion one easily obtains the remainder of multiplication table for all elements of group D_{4h}. Characters of single-valued and double-valued (symbols with prime) of group D_{4h} are presented in Table 4.9.

We consider two space groups $P4/nmm$ and $I4/mmm$ with D_{4h} point group symmetry. Brillouin zones are presented in Fig. 4.5a, b, respectively. Possible pair

Table 4.9 Character table of group D_{4h}. In the first line, digital notations of Kovalev [60] are presented. Symbols with primes correspond to double-valued IRs. (In addition there are also two double-valued IRs, whose characters are multiplied by -1 for the elements with space inversion)

h_i	1	4	2	3	14	15	13	16	25	28	26	27	38	39	37	40
h_i	E	C_{2z}	C_{2x}	C_{2y}	C_{4z}	C_{4z}^3	C_{2b}	C_{2a}	I	σ_z	σ_x	σ_y	S_4	S_4^3	σ_b	σ_a
A_{1g}	1	1	1	1	1	1	1	1	1	1	1	1	1	1	1	1
A_{1u}	1	1	1	1	1	1	1	1	-1	-1	-1	-1	-1	-1	-1	-1
A_{2g}	1	1	-1	-1	1	1	-1	-1	1	1	-1	-1	1	1	-1	-1
A_{2u}	1	1	-1	-1	1	1	-1	-1	-1	-1	1	1	-1	-1	1	1
B_{1g}	1	1	1	1	-1	-1	-1	-1	1	1	1	1	-1	-1	-1	-1
B_{1u}	1	1	1	1	-1	-1	-1	-1	-1	-1	-1	-1	1	1	1	1
B_{2g}	1	1	-1	-1	-1	-1	1	1	1	1	-1	-1	-1	-1	1	1
B_{2u}	1	1	-1	-1	-1	-1	1	1	-1	-1	1	1	1	1	-1	-1
E_g	2	-2	0	0	0	0	0	0	2	-2	0	0	0	0	0	0
E_u	2	-2	0	0	0	0	0	0	-2	2	0	0	0	0	0	0
E_1'	2	0	0	0	$-\sqrt{2}$	$\sqrt{2}$	0	0	2	0	0	0	$-\sqrt{2}$	$\sqrt{2}$	0	0
E_2'	2	0	0	0	$\sqrt{2}$	$-\sqrt{2}$	0	0	2	0	0	0	$\sqrt{2}$	$-\sqrt{2}$	0	0

states for points inside a BZ are the same for these two space groups. However, the results are different for symmetry points on some BZ faces.

The decomposition of Kronecker square into double cosets corresponds to a direct product of wavevector stars $\{k\} \times \{k\}$ of one-electron state. As an example all terms of the square of IR in direction Δ, obtained as direct product of stars, are shown in

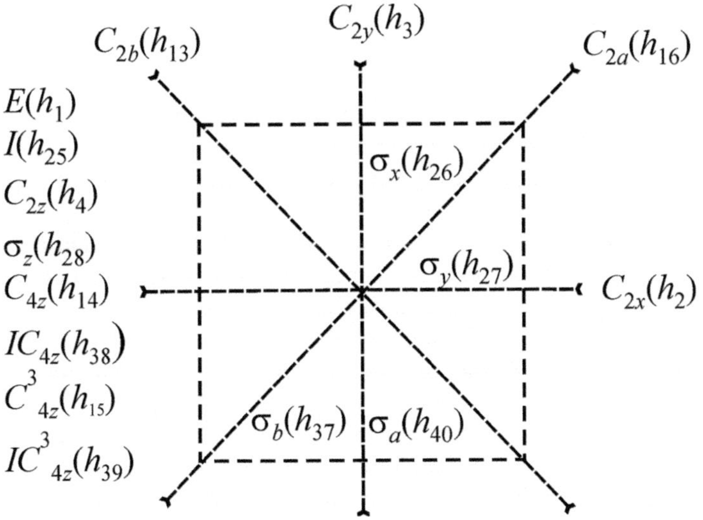

Fig. 4.4 Correspondence between standard and digital notations for the elements of group D_{4h}

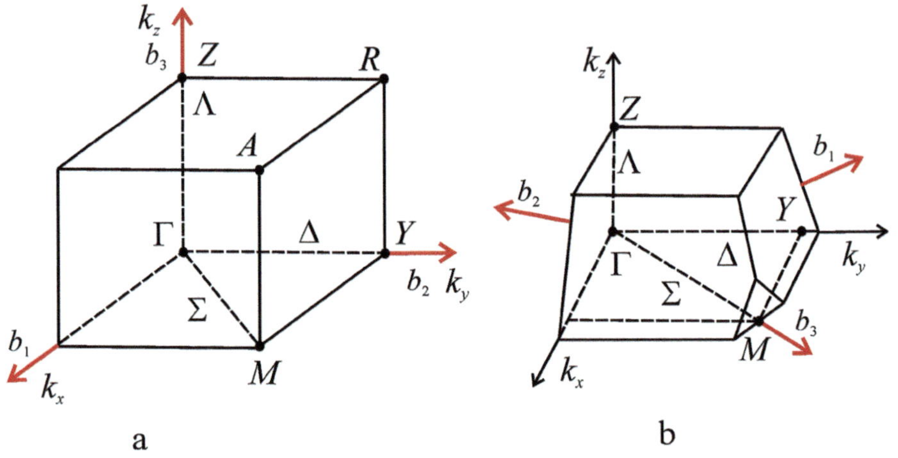

Fig. 4.5 Brillouin zones for the space groups a $P4/nmm$ and b $I4/mmm$

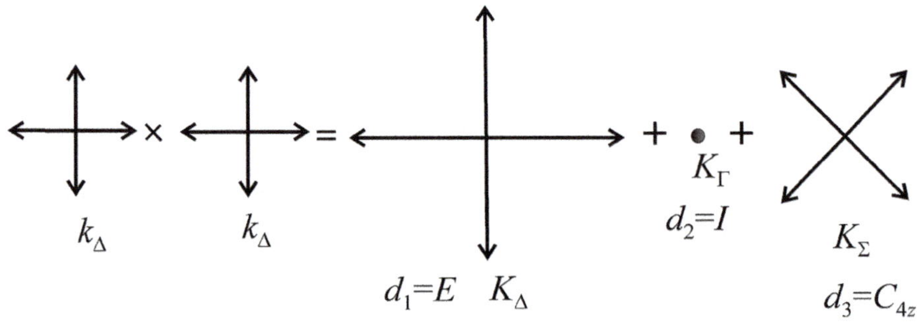

Fig. 4.6 All terms of the decomposition of direct product in direction k_{Δ} (group D_{4h})

Fig. 4.6. There are three types of stars of total wavevector K of a pair, namely, K_{Δ}, K_{Γ} and K_{Σ}. Since it is usually assumed that total momentum of a Cooper pair equals zero, we perform calculation for star at point Γ only. Such a pair is totally symmetric with respect to the lattice translations.

The wavevector groups H and groups $\tilde{M}_{\sigma}$ for some planes and directions are presented in Table 4.10. In planes and directions, the double coset defined by space inversion is considered. In some symmetrical points, space inversion belongs to H and double coset defined by the identity element corresponds to a Cooper pair. In this case, two electrons with the same momentums form a pair and total momentum K of two-electron state equals zero due to translation periodicity.

If k is in one of the symmetry planes (001), (100), (110) and directions Δ, Σ and Λ, the inversion is not included in the wavevector group. In these cases, the group of the wavevector H is extended by the inversion to the group $\tilde{M}_{\sigma} = H + IH$ and symmetrized P^{+} and antisymmetrized P^{-} squares are constructed on this group, which are then induced into the entire point group $\bar{G}$. Using the Frobenius reciprocity

Table 4.10 Wavevector groups H and their extensions $\tilde{M}_\alpha$ for planes and directions in a BZ zone for the space group $P4/nmm$ (see Fig. 4.5)

Location in a BZ	H	$\tilde{M}_\sigma$
Planes (100), (110), (001)	C_s	C_{2h}
Directions Δ, Σ	C_{2v}	D_{2h}
Direction Λ	C_{4v}	D_{4h}
Points R, Y	D_{2h}	D_{2h}
Points Γ, Z, M, A	D_{4h}	D_{4h}

Table 4.11 Characters of representations $P^\pm$ constructed by Mackey theorem and symmetry of singlet two-electron states in directions Δ and Σ (see Fig. 4.5)

									$P^+ \uparrow G$	$P^- \uparrow G$
Δ	h_1	h_2	h_{27}	h_{28}	h_{25}	h_{26}	h_3	h_4	$A_{1g} + B_{1g}$	E_u
Σ	h_1	h_{16}	h_{28}	h_{37}	h_{25}	h_{40}	h_4	h_{13}	$A_{1g} + B_{2g}$	E_u
P^+	1	1	1	1	1	1	1	1		
P^-	1	1	1	1	-1	-1	-1	-1		

theorem (4.2.15), one easily obtains the frequency of appearance of the IR Γ^q of $\bar{G}$ in the decomposition of the representations induced from $\tilde{M}_\sigma$.

The wavevector groups in directions Δ and Σ are isomorphic and the results for these two directions are presented in Table 4.11.

Possible symmetries of spatial parts of even and odd two-electron states on planes of symmetry are presented in Table 4.12. Consider for example odd spatial parts (triplet case). In the basal plane (001) one-dimensional odd IRs are forbidden. This means symmetry protected lines of nodes for IRs A_{1u}, A_{2u}, B_{1u} and B_{2u}. In vertical planes (100) and (010) IR B_{1u} is forbidden, but in vertical planes (−110) and (110) IR B_{2u} is forbidden resulting in corresponding nodes. However, the nodes of E_u on vertical planes are undefined. Indeed, it follows from Mackey theorem that for k a general point in a BZ the number of appearance of any odd (even) IR in the decomposition of antisymmetrized (symmetrized) square equals to its dimension. Thus at general $k-$point E_u appears twice, but on vertical planes just ones. Hence, it follows that one basis set of E_u symmetry vanishes at planes. It will be shown in following sections that these basis sets may by labeled by additional quantum numbers and constructed by two-step induction via intermediate group.

In the direction Λ (see Fig. 4.5) $H = C_{4v}$ and its extension by I is the entire group D_{4h} and additional induction (see formulas (4.2.12) and (4.2.13) is not required. The results for all small one-dimensional IRs coincide: for singlet pairs A_{1g} is only possible and for triplet pairs A_{2u} is only possible (see Table 4.13). In the case of a two-dimensional IR E, the symmetrized square contains three even IRs and one odd IR, while the antisymmetrized one contains three odd IRs and one even IR, and all IRs are one-dimensional. Thus, the calculation confirmed the assertion proved in Sect.

Table 4.12 Characters of representations $P^{\pm}$ constructed by Mackey theorem and possible IRs of two-electron states on symmetry planes

	E	σ	I	C_2	$P^+ \uparrow G$	$P^- \uparrow G$
ΓZY (100), (010)	h_1	h_{26}	h_{25}	h_2	$A_{1g} + B_{1g} + E_g$	$A_{2u} + B_{2u} + E_u$
ΓYM (001)	h_1	h_{28}	h_{25}	h_4	$A_{1g} + A_{2g} + B_{1g} + B_{2g}$	$2E_u$
ΓZM (110), (−110)	h_1	h_{37}	h_{25}	h_{13}	$A_{1g} + B_{2g} + E_g$	$A_{2u} + B_{1u} + E_u$
P^+	1	1	1	1		
P^-	1	1	−1	−1		

Table 4.13 Characters of representations $P^{\pm}$ constructed by Mackey theorem for direction Λ (see Fig. 4.5) and their decomposition. The results for all one-dimensional IRs of C_{4v} (A_1, A_2, B_1 and B_2) are the same. In the last line E_1' is a double-valued IR. Asterisk denotes cases where additional factor −1 appears for double-valued IRs. The result for the other double-valued IR E_2' is the same as for E_1' (the representations of wavevector group C_{4v} are given in Table 1.1)

h_i	1	4	14	15	26	27	37	40	25	28	38	39	2	3	13	16	
$P_{A_1}^+$	1	1	1	1	1	1	1	1	1	1	1	1	1	1	1	1	A_{1g}
$P_{A_1}^-$	1	1	1	1	1	1	1	1	−1	−1	−1	−1	−1	−1	−1	−1	A_{2u}
P_E^+	4	4	0	0	0	0	0	0	2	2	−2	−2	2	2	2	2	$A_{1g}+B_{1g}+B_{2g}+A_{1u}$
P_E^-	4	4	0	0	0	0	0	0	−2	−2	2	2	−2	−2	−2	−2	$A_{2u}+B_{1u}+B_{2u}+A_{2g}$
$P_{E_1'}^-$	4	0	2	2	0	0	0	0	−2	2*	−2	2*	2*	2*	2*	2*	$A_{1u}+E_u+A_{1g}$

4.2 that there is a violation of the direct relationship between the multiplicity and parity of the pair in the case of two-dimensional small IRs. In the case of strong spin-orbit coupling, one-electron states in k_z−direction belong to double-valued IRs E_1' and E_2' of group C_{4v} (see Table 4.9). The results for these two IRs are the same and in Table 4.13 only $P_{E_1'}^-$ is presented. It should be noted that in Mackey theorem the term $D(amam)$ is obtained from the relation $D(am)D(am) = D(amam)$ and in the case of double-valued IRs it should be written as $D(am)D(am) = D(amam)\omega(am, am)$. The cases when such factors appear are denoted in Table 4.8 by the asterisk. Since in our case $a = I$, it is clear that $\omega(Im, Im) = \omega(m, m)$. Those cases in which the sign has been changed due to this factor are marked in Table 4.13 with an asterisk.

At the point Z on the BZ boundary, the wavevector group H coincides with the full group G and the squared IRs are symmetrized (antisymmetrized) according to the standard formulas for point groups:

$$\chi\left[D(g) \times D(g)\right] = \frac{1}{2}\left\{\chi^2\left[D(g)\right] + \chi\left[D(g^2)\right]\omega\left(g.g\right)\right\} \quad (4.5.2)$$

$$\chi\left\{D(g) \times D(g)\right\} = \frac{1}{2}\left\{\chi^2\left[D(g)\right] - \chi\left[D(g^2)\right]\omega\left(g.g\right)\right\} \quad (4.5.3)$$

Table 4.14 Characters of representations $P^{\pm}$ constructed by Mackey theorem for point Z for symmorphic space group $P4/mmm$ (see Fig. 4.5). Double-valued IR is denoted by prime

	1	4	2	3	14	15	13	16	25	28	26	27	38	39	37	40	IR
$[A_i^2], [B_i^2]$	1	1	1	1	1	1	1	1	1	1	1	1	1	1	1	1	A_{1g}
$\{A_i^2\}, \{B_i^2\}$	0	0	0	0	0	0	0	0	0	0	0	0	0	0	0	0	No
$[E_i^2]$	3	3	1	1	−1	−1	1	1	3	3	1	1	−1	−1	1	1	$A_{1g}+B_{1g}+B_{2g}$
$\{E_i^2\}$	1	1	−1	−1	1	1	−1	−1	1	1	−1	−1	1	1	−1	−1	A_{2g}
$\{E_i'^2\}$	1	1	1	1	1	1	1	1	1	1	1	1	1	1	1	1	A_{1g}

In this case, all IRs for both singlet pair and the triplet pairs are even, and the antisymmetrized square of the one-dimensional IR is equal to zero. This means that for one-dimensional small IR triplet pairs are forbidden. The decomposition is shown in Table 4.14. The decompositions of all antisymmetrized double-valued IRs include A_{1g} and A_{2g} only.

4.6 Non-symmorphic Space Groups, Symmetry Points in a BZ

In non-symmorphic space groups, projective IRs are possible on the BZ boundary, so in addition to even IRs, considered in the previous section, odd IRs can also appear in the double coset, defined by the identity element. Consider, as an example, the point M $(b_1/2, b_2/2, 0)$ in the BZ for the space group $P4/nmm$ (D_{4h}^7) (see Fig. 4.5 a), where the condition that the resulting vector equals to zero has the form:

$$k_M + k_M - b_M = 0 \tag{4.6.1}$$

Thus symmetrized and antisymmetrized squares are calculated according to the formulas (4.5.2) and (4.5.3). To perform these calculations, we need projective factors for the squares of elements. In group D_{4h}^7, the elements h_2, h_3, h_{38}, h_{39}, h_{13}, h_{25}, h_{16}, h_{28} are connected with an improper translation $\tau = (1/2, 1/2, 0)$ [60]. Using the definition of group elements shown in Fig. 4.4, we can easily define that the elements h_2, h_3, h_{38}, h_{39} transform the vector $(b_1/2, b_2/2, 0)$ into vectors $(\pm b_1/2, \mp b_2/2, 0)$ and vector $k_M - h_j k_M$ equals to b_1 or to b_2. Thus for the squares of the elements $\{h_j|\tau_j\}$ $(j = 2, 3, 38, 39)$ the phase factors $\omega(h_j, h_j) = \exp(i(k_M - h_j^{-1} k_M) \cdot \tau_j)$ in formulas (4.5.2) and (4.5.3) equal to -1. The elements h_{13}, h_{25} translate the vector $(b_1/2, b_2/2, 0)$ into $(-b_1/2, -b_2/2, 0)$ position and vector $k_M - h_j k_M$ equals to $(b_1, b_2, 0)$. By this means we obtain that the projective factors for the squares of h_{13} and h_{25} equal to 1. Also, since the elements h_{16} and h_{28} don't change the wavevector k_M, the projective factors for the squares of these elements equal to 1. Projective representation for point M in a BZ of group D_{4h}^7, the phase factors for the squares of the elements and the symmetrized and antisymmetrized squares are presented in Table 4.15.

Table 4.15 Characters of single-valued projective representations $\hat{t}_i$ at the point M in a BZ for the space group D_{4h}^7(IR T_{154} [60]), projective factors $\omega(h_i^2)$ and symmetrized and antisymmetrized squares

h_i	E	h_4	h_2	h_3	h_{14}	h_{15}	h_{13}	h_{16}	h_{25}	h_{28}	h_{26}	h_{27}	h_{38}	h_{39}	h_{37}	h_{40}
h_i^2	h_1	h_1	h_1	h_1	h_4	h_4	h_1	h_1	h_1	h_1	h_1	h_1	h_4	h_4	h_1	h_1
$\omega(h_i^2)$	1	1	−1	−1	1	1	1	1	1	1	1	1	−1	−1	1	1
$\hat{t}_1$	2	−2	0	0	0	0	−2	2	0	0	0	0	0	0	0	0
$\hat{t}_2$	2	−2	0	0	0	0	2	−2	0	0	0	0	0	0	0	0
$\hat{t}_3$	2	2	0	0	0	0	0	0	0	0	0	0	0	0	−2	−2
$\hat{t}_4$	2	2	0	0	0	0	0	0	0	0	0	0	0	0	2	2
$\left[\hat{t}_1^2\right]$	3	3	−1	−1	−1	−1	3	3	1	1	1	1	1	1	1	1
$\left\{\hat{t}_1^2\right\}$	1	1	1	1	1	1	1	1	−1	−1	−1	−1	−1	−1	−1	−1
$\left[\hat{t}_3^2\right]$	3	3	−1	−1	1	1	1	1	1	1	1	1	−1	−1	3	3
$\left\{\hat{t}_3^2\right\}$	1	1	1	1	−1	−1	−1	−1	−1	−1	−1	−1	1	1	1	1

The decompositions of symmetrized and antisymmetrized squares for point M in a BZ of non-symmorphic space group have some remarkable features.

The expansion of symmetrized squares of projective IRs t_i at the point M gives two even IRs and one odd IR:

$$\left[t_1^2\right] = \left[t_2^2\right] = A_{1g} + B_{2g} + B_{2u}$$

$$\left[t_3^2\right] = \left[t_4^2\right] = A_{1g} + B_{2g} + A_{2u}$$

Antisymmetrized squares are odd in all cases:

$$\left\{t_1^2\right\} = \left\{t_2^2\right\} = A_{1u}$$

$$\left\{t_3^2\right\} = \left\{t_4^2\right\} = B_{1u}$$

Thus, at points of the BZ, where the inversion belongs to the wavevector group, there is a violation of the unambiguous relationship between the multiplicity and parity of the pair. Moreover, in the case of one-dimensional IRs and symmorphic groups, the character of the antisymmetrized square is equal to zero, and triplet pairs are forbidden.

4.7 Spatial Part of Cooper Pair Wavefunction

In a general point of a BZ, the wavevector group H consists of an identity element only and the expressions for spatial parts of singlet and triplet pairs which follow from Mackey theorem are the same as that follows from Anderson ansatz [78]:

$$\psi_{k_1}^{s(t)} = \varphi_{k_1}\varphi_{Ik_1} \pm \varphi_{Ik_1}\varphi_{k_1} \tag{4.7.1}$$

where k_1 denotes a general point in a representation domain of a BZ and the number of electronic coordinate is the number of position in a product. Two-electron functions in formula (4.7.1) with plus (minus) sign are even (odd) with respect to permutation of electron coordinates. Hence, it follows that the function with plus (minus) sign corresponds to singlet (triplet) state.

The action of the left coset representatives s_σ (4.2.1) on k_1 results in a star $\{k_1\}$. The number of one-electron basis functions for k_1 a general point in a BZ equals to $|\bar{G}|$. All one-electron wavefunctions are obtained when k_1 runs over basis (representation) domain of a BZ, whose volume is $1/|\bar{G}|$ of the whole BZ volume [19]. According to the space-group approach pair functions are constructed for all k_1 in the representation domain of a BZ making use of formula (4.7.1). Since spatial inversion is already included in the definition of a pair (4.7.1), the number of elements in the basis of the direct product is $|\bar{G}|/2$ for singlet pairs and $|\bar{G}|/2$ for triplet pairs. This relationship applies only to two-electron states with zero total momentum, but total number of two-electron states equals to $|\bar{G}|^2$. These two $|\bar{G}|/2$–dimensional point group basis sets are reducible and may be decomposed into irreducible parts. The basis set of IR Γ^q of point group $\bar{G}$ may be obtained making use of standard projection operator formula:

$$\Psi_i^{\Gamma^q} = \frac{|\Gamma^q|}{|\bar{G}'|} \sum_{h_j \in \bar{G}'} \Gamma_{ii}^q\left(h_j\right) h_j \psi_{k1}^{s(t)} \tag{4.7.2}$$

where $\bar{G}'$ is any subgroup of index 2 of $\bar{G}$, which does not contain inversion I, but it may contain mirror reflections. It is clear from formula (4.7.1) that singlet (triplet) pairs are even (odd) with respect to space inversion. Hence, it follows that in formula (4.7.2) singlet and triplet functions are projected on even and odd IRs, respectively.

In the case of D_{4h} point group, it is convenient to choose C_{4v} as a halving subgroup $\bar{G}'$. Basis set for singlet (triplet) pair for k a general point in a BZ is shown in Fig. 4.7 where subscript k and superscripts s or t dropped.

Basis functions shown in Fig. 4.7 form the basis of induced representations $A_{g(Ci)} \uparrow D_{4h}$ or $A_{u(C_i)} \uparrow D_{4h}$ in singlet and triplet cases, respectively. The action of all elements of halving subgroup $\bar{G}'$ results in permutations of the elements of each set (singlet or triplet). The other elements of $\bar{G}$ may be represented as products $I\bar{G}'$ and their action results in change of the sign of triplet functions an addition to permutations. The results of the action of group elements on these basis functions may be obtained making use of multiplication table of the group D_{4h}, presented in Table 4.8 and formula (4.5.1) as follows. Let us introduce $|\bar{G}'|/2$ auxiliary functions,

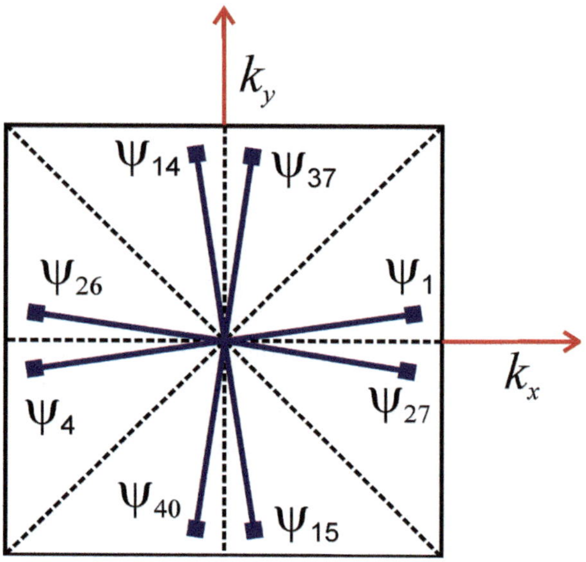

Fig. 4.7 Cooper pair basis set in D_{4h} symmetry. The functions are obtained by the action of the elements of group C_{4v} on the functions defined by formula (4.7.1)

connected with the $\left|\bar{G}'\right|/2$ basis functions (see formula (4.7.1) by the space inversion $I = h_{25}$. These additional functions are connected with the basis functions by the following relations:

$$h_{25}\psi_i^{s(t)} = \begin{cases} \pm\psi_{i+24}^{s(t)}, & \text{if } 1 \le i \le 24 \\ \pm\psi^{s(t)}{}_{i-24} & \text{if } 25 \le i \le 48 \end{cases} \tag{4.7.3}$$

where plus and minus signs correspond to singlet and triplet cases, respectively. Thus we distinguish two functions $\psi_{27}^t = (\varphi_{27}\varphi_3 - \varphi_3\varphi_{27})$ and $\psi_3^t = (\varphi_3\varphi_{27} - \varphi_{27}\varphi_3)$. We can write ψ_k for $h_i\psi_j$ if it follows from the multiplication table of the group that $h_k = h_i h_j$. If the resulting function is not in the list of $\left|\bar{G}\right|/2$ basis functions it should be transformed making use of relations (4.7.3).

Singlet and triplet pair functions belonging to one-dimensional IRs constructed making use of formula (4.7.2) and their nodal planes are presented in Tables 4.16 and 4.17, respectively. Nodal planes are shown in Figs. 4.8 and 4.9, where nodes in vertical planes are represented by red lines and nodes in basal plane by big red circles. Nodal planes of one-dimensional IRs uniquely determined by the characters of representations. For example, characters of B_{1g} equal to $+1$ for reflections in all coordinate planes and equal to -1 for reflections in vertical diagonal planes, resulting in nodes in these planes (see Fig. 4.8). In B_{1u}, the characters for reflection elements are obtained multiplying characters of B_{1g} by -1 and nodal and nodeless planes are interchanged with respect to B_{1g}. Starting with any k_1 in the sector $0 < k_1 < \pi/4$, the signs (relative phases) in all other sectors are obtained group-theoretically. However, there is one more free parameter—phase winding in the first sector. This phase winding will be considered in subsequent sections.

Table 4.16 Functions of singlet pairs and their nodal planes in D_{4h} symmetry. Basis functions ψ_i^s are shown in Fig. 4.7

IR	Singlet function	Nodal planes
$\Psi(A_{1g})$	$\psi_1^s+\psi_4^s+\psi_{14}^s+\psi_{15}^s+\psi_{26}^s+\psi_{27}^s+\psi_{37}^s+\psi_{40}^s$	No
$\Psi(A_{2g})$	$\psi_1^s+\psi_4^s+\psi_{14}^s+\psi_{15}^s-\psi_{26}^s-\psi_{27}^s-\psi_{37}^s-\psi_{40}^s$	(100), (010), (110), (−110)
$\Psi(B_{1g})$	$\psi_1^s+\psi_4^s-\psi_{14}^s-\psi_{15}^s+\psi_{26}^s+\psi_{27}^s-\psi_{37}^s-\psi_{40}^s$	(110) , (−110)
$\Psi(B_{2g})$	$\psi_1^s+\psi_4^s-\psi_{14}^s-\psi_{15}^s-\psi_{26}^s-\psi_{27}^s+\psi_{37}^s+\psi_{40}^s$	(100), (010)

Table 4.17 Spatial parts of triplet pairs and their nodal planes in D_{4h} symmetry. Basis functions ψ_i^s are shown in Fig. 4.7

IR	Triplet function	Nodal planes
$\Psi(A_{1u})$	$\psi_1^t+\psi_4^t+\psi_{14}^t+\psi_{15}^t-\psi_{26}^t-\psi_{27}^t-\psi_{37}^t-\psi_{40}^t$	All planes
$\Psi(A_{2u})$	$\psi_1^t+\psi_4^t+\psi_{14}^t+\psi_{15}^t+\psi_{26}^t+\psi_{27}^t+\psi_{37}^t+\psi_{40}^t$	(001)
$\Psi(B_{1u})$	$\psi_1^t+\psi_4^t-\psi_{14}^t-\psi_{15}^t-\psi_{26}^t-\psi_{27}^t+\psi_{37}^t+\psi_{40}^t$	(100) , (010), (001)
$\Psi(B_{2u})$	$\psi_1^t+\psi_4^t-\psi_{14}^t-\psi_{15}^t+\psi_{26}^t+\psi_{27}^t-\psi_{37}^t-\psi_{40}^t$	(110) , (−110), (001)

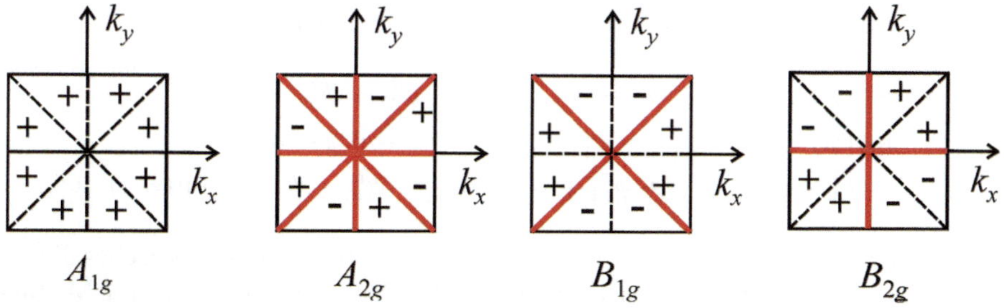

Fig. 4.8 Signs and nodes (red lines) of basis functions of one-dimensional even IRs of D_{4h} group in the plane normal to k_z. Basal plane $k_z = 0$ is nodeless for all IRs

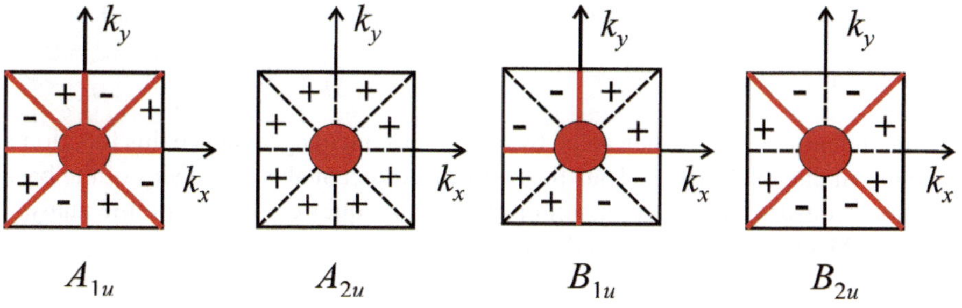

Fig. 4.9 Signs and nodes (red lines) of basis functions of one-dimensional odd IRs of D_{4h} group in the plane normal to k_z. Big red circles denote nodes in the basal plane

These nodal structures are the same as obtained in Table 4.12 by Mackey and Frobenius reciprocity theorems. Nodal structure can be also verified directly by considering the action by reflections in planes on these functions. Consider as an example nodal structure of $\Psi(B_{1u})$. Making use of group multiplication rules is presented in Table 4.8 and formulas (4.5.1) and (4.7.3) one easily obtains the result of the action of h_{27} (reflection in plane (010) on the function $\Psi(B_{1u})$:

$$\sigma_y \Psi(B_{1u}) = h_{27}\ \left(\psi_1^t + \psi_4^t - \psi_{14}^t - \psi_{15}^t - \psi_{26}^t - \psi_{27}^t + \psi_{37}^t + \psi_{40}^t\right) =$$
$$\psi_{27}^t + \psi_{26}^t - \psi_{40}^t\ - \psi_{37}^t - \psi_4^t - \psi_1^t + \psi_{15}^t + \psi_{14}^t = -\Psi(B_{1u})$$

Suppose that the wavevector k_1 (and function ψ_1 in Fig. 4.7) approaches the plane (010). In this case, the vector $h_{27}k_1$ (and function ψ_{27} in Fig. 4.7) approaches this plane. On the plane they can merge or cancel. In is seen in Fig. 4.7 that the same occurs with pairs of functions (ψ_{14}, ψ_{37}), (ψ_4, ψ_{26}) and (ψ_{15}, ψ_{40}) on equivalent planes. Since we obtained that reflection σ_y changes the sign of $\Psi(B_{1u})$ on the plane (010), we conclude that this function has nodes in this plane. Similarly one obtains that there are nodes of $\Psi(B_{1u})$ in the equivalent plane (100).

The reflection in basal plane (001) doesn't belong to group $\bar{G}' = C_{4v}$ and it is convenient to represent it as a product of space inversion h_{25} and rotation h_4 about $z-$axis by π, and take into account oddness of triplet functions:

$$h_{25}h_4\ \left(\psi_1^t + \psi_4^t - \psi_{14}^t - \psi_{15}^t - \psi_{26}^t - \psi_{27}^t + \psi_{37}^t + \psi_{40}^t\right) =$$
$$h_{25}\left(\psi_4^t + \psi_1^t - \psi_{15}^t\ - \psi_{14}^t - \psi_{27}^t - \psi_{26}^t + \psi_{40}^t + \psi_{37}^t\right) =$$
$$-\psi_4^t - \psi_1^t + \psi_{15}^t\ + \psi_{14}^t + \psi_{27}^t + \psi_{26}^t - \psi_{40}^t - \psi_{37}^t = -\Psi(B_{1u})$$

Thus the function $\Psi(B_{1u})$ has nodes in basal plane. If vector k_1 approaches diagonal plane (110), pairs of function are connected by plane reflection $\sigma_a = h_{40}$. Making use of group multiplication table one obtains that the function remains unchanged under the action of h_{40}:

$$\sigma_a \Psi(B_{1u}) = h_{40}\ \left(\psi_1^t + \psi_4^t - \psi_{14}^t - \psi_{15}^t - \psi_{26}^t - \psi_{27}^t + \psi_{37}^t + \psi_{40}^t\right) =$$
$$\psi_{40}^t + \psi_{37}^t - \psi_{26}^t\ - \psi_{27}^t - \psi_{14}^t - \psi_{15}^t + \psi_4^t + \psi_1^t = \Psi(B_{1u})$$

The function $\Psi(B_{1u})$ when acted by σ_a remains unchanged, hence this function is nodeless in diagonal planes. Thus the results of direct analysis of nodes of the basis functions constructed by projection operator techniques coincide with the analysis based on Mackey and Frobenius theorems.

4.8 Total Triplet Pair Functions in $L - S$ Coupling

For k a general point of a BZ, a complete basis set of triplet pairs consists of three functions, differing by their spin projections:

$$\psi^t_{\hat{x}} = \hat{x} \left(\varphi_k \varphi_{Ik} - \varphi_{Ik} \varphi_k\right) \tag{4.8.1}$$

$$\psi^t_{\hat{y}} = \hat{y} \left(\varphi_k \varphi_{Ik} - \varphi_{Ik} \varphi_k\right) \tag{4.8.2}$$

$$\psi^t_{\hat{z}} = \hat{z} \left(\varphi_k \varphi_{Ik} - \varphi_{Ik} \varphi_k\right) \tag{4.8.3}$$

Thus total dimension of the space of triplet two-electron functions in D_{4h} symmetry equals to 24. Following theorem on the product of IR Σ^i of the entire group G and a representation $D^k \uparrow G$ induced from a subgroup H is valid [19, 20]:

$$\Sigma^i \times \left(D^k \uparrow G\right) = \left(\left(\Sigma^i \downarrow H\right) \times D^k\right) \uparrow G \tag{4.8.4}$$

This means that there are two equivalent ways to construct a total pair function. One can construct a spatial part $D^k \uparrow G$ in entire group and then decompose a direct product of spatial part and spin part Σ^i. Also one can reduce representation of triplet spin Σ^i on the subgroup H, construct direct products of spatial and spin parts, and then induce these spin orbitals into G. For k a general point of a BZ the induction and further decompositions in IRs are equivalent to the use of standard projection operator technique [2]. Consider axial symmetry groups and take D_{4h} as an example. In D_{4h} symmetry two triplet ESP (equal spin pairing) components $(\hat{x}, \hat{y})$ belong to E_g and OSP (opposite spin pairing) component $\hat{z}$ belongs to A_{2g}. Multiplication of spatial parts by OSP function $\hat{z}$ results in phase factors for reflections in vertical planes and for rotations in horizontal axes, but don't change the structure of the spatial function, i.e. one-dimensional IRs A_{1u}, A_{2u}, B_{1u} and B_{2u} remain one-dimensional and two-dimensional IR E_u remains two-dimensional. Since the product $E_g \times E_u$ is decomposed into $A_{1u} + A_{2u} + B_{1u} + B_{2u}$ there are four nontrivial ESP combinations called helical. Note that because E_u appears twice in the decomposition of triplet spatial basis, there are eight distinct helical states.

Triplet basis functions are obtained by the action of projection operators on spin orbitals:

$$\Psi^{\Gamma^q}_{(\alpha)i} = \frac{|\Gamma^q|}{|\bar{G}'|} \sum_{h_j \in \bar{G}'} \Gamma^q_{ii}\left(h_j\right) h_j \hat{r}_\alpha \psi^t_{k1} \tag{4.8.5}$$

where sum runs over the halving subgroup $\bar{G}'$ and $\hat{r}_\alpha$ stands for one of the α triplet spin components. In formula (4.8.5), it is assumed that the same group elements h_j act in $k-$space and in spin space. The basis set of a Cooper pair with initial spin $\hat{x}$ is shown in Fig. 4.10. Similar basis sets can be constructed with initial spin functions $\hat{y}\psi_1$ and $\hat{z}\psi_1$.

Fig. 4.10 Basis of Cooper pair real ESP spin orbitals in D_{4h} symmetry with initial spin $\hat{x}$

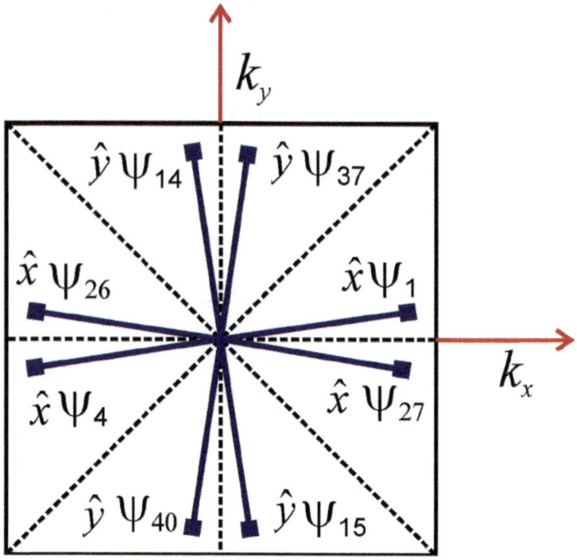

Triplet basis functions of one-dimensional IRs obtained making use of formula (4.8.5) are presented in Table 4.18.

Linear combinations of Table 4.18 are constructed for an arbitrary k_1 in basis domain of a BZ, i.e. in sector $0 \le \theta \le \pi/4$ in Fig. 4.10. When k_1 approaches k_x-axis (plane (010) in three dimensions) functions with the same spin projections, namely, $(\hat{x}\psi_1^t, \hat{x}\psi_{27}^t)$, $(\hat{x}\psi_4^t, \hat{x}\psi_{26}^t)$, $(\hat{y}\psi_{14}^t, \hat{y}\psi_{37}^t)$ and $(\hat{y}\psi_{15}^t, \hat{y}\psi_{40}^t)$ meet at vertical coordinate planes. It is assumed that the wavevector k_1 has nonzero k_z-component. When k_z approaches zero pairs of functions $(\hat{x}\psi_1^t, \hat{x}\psi_4^t)$, $(\hat{x}\psi_{26}^t, \hat{x}\psi_{27}^t)$, $(\hat{y}\psi_{14}^t, \hat{y}\psi_{15}^t)$ and $(\hat{y}\psi_{37}^t, \hat{y}\psi_{40}^t)$ meet at basal plane. On the planes pairs of functions can merge or cancel. If total function changes a sign under the reflection in plane it has a node on the plane. If total function is invariant under the action of plane reflection it is nodeless on this plane. Thus nodal structure can be obtained by the analysis of the result of plane reflection as follows. The basis functions in Table 4.18 are numbered with indices of C_{4v} group in notation of [60], which are shown in Fig. 4.4. According to the basis choice, the action of elements of group C_{4v} results in permutations of basis functions without change of the sign, but the action of space inversion $h_{25} = I$ results in change of the sign of spatial part of a triplet pair. Multiplication table for the pure rotation subgroup of D_{4h} (i.e. group D_4) is presented in Table 4.8. Note that the transformations of spin components correspond to an axial vector, i.e. they are invariant with respect to inversion, but under the action of pure rotations they are transformed as vectors.

When acting by h_{26}—a reflection in (100) plane on the function $\Psi_{\hat{x}}(A_{1u})$ we have

$$h_{26}\Psi_{\hat{x}}(A_{1u}) = \hat{x}\left(\psi_{26}^t - \psi_{27}^t - \psi_1^t + \psi_4^t\right) + (-\hat{y})\left(\psi_{37}^t - \psi_{40}^t + \psi_{14}^t - \psi_{15}^t\right) = -\Psi_{\hat{x}}(A_{1u})$$

Table 4.18 Total triplet two-electron states for group D_{4h} and their group-theoretical nodes on planes[1)]

IR	Total wavefunction	Nodes on planes
$\Psi_{\hat{x}}(A_{1u})$	$\hat{x}\left(\psi_1^t-\psi_4^t-\psi_{26}^t+\psi_{27}^t\right)+$ $\hat{y}\left(\psi_{14}^t-\psi_{15}^t+\psi_{37}^t-\psi_{40}^t\right)$	(100), (010), (001)
$\Psi_{\hat{y}}(A_{1u})$	$\hat{y}\left(\psi_1^t-\psi_4^t+\psi_{26}^t-\psi_{27}^t\right)+$ $\hat{x}\left(-\psi_{14}^t+\psi_{15}^t+\psi_{37}^t-\psi_{40}^t\right)$	(100), (010), (001)
$\Psi_{\hat{z}}(A_{1u})$	$\hat{z}\left(\psi_1^t+\psi_4^t+\psi_{26}^t+\psi_{27}^t+\psi_{14}^t+\right.$ $\left.\psi_{15}^t+\psi_{37}^t+\psi_{40}^t\right)$	(100), (010), (001), (110), (−1, 1, 0)
$\psi_{\hat{x}}^t(A_{2u})$	$\hat{x}\left(\psi_1^t-\psi_4^t+\psi_{26}^t-\psi_{27}^t\right)+$ $\hat{y}\left(\psi_{14}^t-\psi_{15}^t-\psi_{37}^t+\psi_{40}^t\right)$	(001)
$\Psi_{\hat{y}}(A_{2u})$	$\hat{y}\left(\psi_1^t-\psi_4^t-\psi_{26}^t+\psi_{27}^t\right)+$ $\hat{x}\left(-\psi_{14}^t+\psi_{15}^t-\psi_{37}^t+\psi_{40}^t\right)$	(001)
$\Psi_{\hat{z}}(A_{2u})$	$\hat{z}\left(\psi_1^t+\psi_4^t-\psi_{26}^t-\psi_{27}^t+\psi_{14}^t+\right.$ $\left.\psi_{15}^t-\psi_{37}^t-\psi_{40}^t\right)$	(001)
$\Psi_{\hat{x}}(B_{1u})$	$\hat{x}\left(\psi_1^t-\psi_4^t-\psi_{26}^t+\psi_{27}^t\right)+$ $\hat{y}\left(-\psi_{14}^t+\psi_{15}^t-\psi_{37}^t+\psi_{40}^t\right)$	(100), (010), (001)
$\Psi_{\hat{y}}(B_{1u})$	$\hat{y}\left(\psi_1^t-\psi_4^t+\psi_{26}^t-\psi_{27}^t\right)+$ $\hat{x}\left(\psi_{14}^t-\psi_{15}^t-\psi_{37}^t+\psi_{40}^t\right)$	(100), (010), (001)
$\Psi_{\hat{z}}(B_{1u})$	$\hat{z}\left(\psi_1^t+\psi_4^t+\psi_{26}^t+\psi_{27}^t-\psi_{14}^t-\psi_{15}^t-\right.$ $\left.\psi_{37}^t-\psi_{40}^t\right)$	(100), (010), (001)
$\Psi_{\hat{x}}(B_{2u})$	$\hat{x}\left(\psi_1^t-\psi_4^t+\psi_{26}^t-\psi_{27}^t\right)+$ $\hat{y}\left(-\psi_{14}^t+\psi_{15}^t+\psi_{37}^t-\psi_{40}^t\right)$	(001)
$\Psi_{\hat{y}}(B_{2u})$	$\hat{y}\left(\psi_1^t-\psi_4^t-\psi_{26}^t+\psi_{27}^t\right)+$ $\hat{x}\left(\psi_{14}^t-\psi_{15}^t+\psi_{37}^t-\psi_{40}^t\right)$	(001)
$\Psi_{\hat{z}}(B_{2u})$	$\hat{z}\left(\psi_1^t+\psi_4^t-\psi_{26}^t-\psi_{27}^t-\psi_{14}^t-\psi_{15}^t+\right.$ $\left.\psi_{37}^t+\psi_{40}^t\right)$	(001), (110), (−1, 1, 0)

[1)] It is seen in Fig. 4.10 that different spin functions of $\hat{x}$- and $\hat{y}$-type meet at diagonal planes and in this spin representation there are nodes in diagonal planes

Since the reflection in plane (100) changed the sign of the basis function $\Psi_{\hat{x}}(A_{1u})$, we conclude that this function is nodal in this plane. The reflection h_{27} in plane (010) also changes the sign of this function:

$$h_{27}\Psi_{\hat{x}}(A_{1u})=(-\hat{x})\left(\psi_{27}^t-\psi_{26}^t-\psi_4^t+\psi_1^t\right)+\hat{y}\left(\psi_{40}^t-\psi_{37}^t+\psi_{15}^t-\psi_{14}^t\right)=-\Psi_{\hat{x}}(A_{1u})$$

Thus the function $\Psi_{\hat{x}}(A_{1u})$ is nodal in the plane (010) also. The reflection in the basal plane doesn't belong to C_{4v} and it is convenient to represent it as $\sigma_z=IC_{2z}=Ih_4$ and to explore the oddness of triplet function:

$$Ih_4\Psi_{\hat{x}}(A_{1u})=I(-\hat{x})\left(\psi_4^t-\psi_1^t-\psi_{27}^t+\psi_{26}^t\right)+I(-\hat{y})\left(\psi_{15}^t-\psi_{14}^t+\psi_{40}^t-\psi_{37}^t\right)=-\psi_{\hat{x}}^t(A_{1u})$$

The reflection in basal plane doesn't change the sign of the function, hence it follows that the function is nodeless in the basal plane.

Also we can begin with projecting $\hat{y}$ spin component and check the nodal structure in this case by a similar way:

$$
\begin{aligned}
&h_{26}\Psi_{\hat{y}}(A_{1u}) = (-\hat{y})\left(\psi_{26}^t - \psi_{27}^t + \psi_1^t - \psi_4^t\right) + \hat{x}\left(-\psi_{37}^t + \psi_{40}^t + \psi_{14}^t - \psi_{15}^t\right) = -\Psi_{\hat{y}}(A_{1u})\\
&h_{27}\Psi_{\hat{y}}(A_{1u}) = \hat{y}\left(\psi_{27}^t - \psi_{26}^t + \psi_4^t - \psi_1^t\right) + (-\hat{x})\left(-\psi_{40}^t + \psi_{37}^t + \psi_{15}^t - \psi_{14}^t\right) = -\Psi_{\hat{y}}(A_{1u})\\
&Ih_4\Psi_{\hat{y}}(A_{1u}) = I(-\hat{y})\left(\psi_4^t - \psi_1^t + \psi_{27}^t - \psi_{26}^t\right) + I\,(-\hat{x})\left(-\psi_{15}^t + \psi_{14}^t + \psi_{40}^t - \psi_{37}^t\right) = -\Psi_{\hat{y}}(A_{1u})
\end{aligned}
$$

Thus function $\Psi_{\hat{y}}(A_{1u})$ is nodal in all coordinate planes:

$$
\begin{aligned}
&h_{26}\Psi_{\hat{z}}(A_{1u}) = (-\hat{z})\left(\psi_{26}^t + \psi_{27}^t + \psi_1^t + \psi_4^t + \psi_{37}^t + \psi_{40}^t + \psi_{14}^t + \psi_{15}^t\right) = -\Psi_{\hat{z}}(A_{1u})\\
&h_{27}\Psi_{\hat{z}}(A_{1u}) = (-\hat{z})\left(\psi_{27}^t + \psi_{26}^t + \psi_4^t + \psi_1^t + \psi_{40}^t + \psi_{37}^t + \psi_{15}^t + \psi_{14}^t\right) - \Psi_{\hat{z}}(A_{1u})\\
&h_{28}\Psi_{\hat{z}}(A_{1u}) = Ih_4\Psi_{\hat{z}}(A_{1u}) = I\hat{z}\left(\psi_4^t + \psi_1^t + \psi_{27}^t + \psi_{26}^t + \psi_{15}^t + \psi_{14}^t + \psi_{40}^t + \psi_{37}^t\right) = -\Psi_{\hat{z}}(A_{1u})
\end{aligned}
$$

Going over to the function of A_{2u} type, we obtain

$$
\begin{aligned}
&h_{26}\Psi_{\hat{x}}(A_{2u}) = \hat{x}\left(\psi_{26}^t - \psi_{27}^t + \psi_1^t - \psi_4^t\right) + (-\hat{y})\left(\psi_{37}^t - \psi_{40}^t - \psi_{14}^t + \psi_{15}^t\right) = \Psi_{\hat{x}}(A_{2u})\\
&h_{27}\Psi_{\hat{x}}(A_{2u}) = (-\hat{x})\left(\psi_{27}^t - \psi_{26}^t + \psi_4^t - \psi_1^t\right) + (\hat{y})\left(\psi_{40}^t - \psi_{37}^t - \psi_{15}^t + \psi_{14}^t\right) = \Psi_{\hat{x}}(A_{2u})\\
&Ih_4\Psi_{\hat{x}}(A_{2u}) = I\,(-\hat{x})\left(\psi_4^t - \psi_1^t + \psi_{27}^t - \psi_{26}^t\right) + I\,(-\hat{y})\left(\psi_{15}^t - \psi_{14}^t - \psi_{40}^t + \psi_{37}^t\right) = -\Psi_{\hat{x}}(A_{2u})\\
&h_{26}\Psi_{\hat{y}}\,(A_{2u}) = (-\hat{y})\left(\psi_{26}^t - \psi_{27}^t - \psi_1^t + \psi_4^t\right) + \hat{x}\left(-\psi_{37}^t + \psi_{40}^t - \psi_{14}^t + \psi_{13}^t\right) = \Psi_{\hat{y}}\,(A_{2u})\\
&h_{27}\Psi_{\hat{y}}\,(A_{2u}) = \ \hat{y}\left(\psi_{27}^t - \psi_{26}^t - \psi_4^t + \psi_1^t\right) + (-\hat{x})\left(-\psi_{40}^t + \psi_{37}^t - \psi_{15}^t + \psi_{40}^t\right) = \Psi_{\hat{y}}\,(A_{2u})\\
&Ih_4\Psi_{\hat{y}}\,(A_{2u}) = \ I\,(-\hat{y})\left(\psi_4^t - \psi_1^t - \psi_{27}^t + \psi_{26}^t\right) + I\,(-\hat{x})\left(-\psi_{15}^t + \psi_{14}^t - \psi_{40}^t + \psi_{37}^t\right) = -\Psi_{\hat{y}}\,(A_{2u})\\
&h_{26}\Psi_{\hat{z}}(A_{2u}) = (-\hat{z})\left(\psi_{26}^t + \psi_{27}^t - \psi_1^t - \psi_4^t + \psi_{37}^t + \psi_{40}^t - \psi_{14}^t - \psi_{15}^t\right) = \Psi_{\hat{z}}(A_{2u})\\
&h_{27}\Psi_{\hat{z}}(A_{2u}) = (-\hat{z})\left(\psi_{27}^t + \psi_{26}^t - \psi_4^t - \psi_1^t + \psi_{40}^t + \psi_{37}^t - \psi_{15}^t - \psi_{14}^t\right) = \Psi_{\hat{z}}(A_{2u})\\
&Ih_4\Psi_{\hat{z}}(A_{2u}) = I\hat{z}\left(\psi_4^t + \psi_1^t - \psi_{27}^t - \psi_{26}^t + \psi_{15}^t + \psi_{14}^t - \psi_{40}^t - \psi_{37}^t\right) = -\Psi_{\hat{z}}(A_{2u})
\end{aligned}
$$

Going over to the function of B_{1u} type, we obtain

$$
\begin{aligned}
&h_{26}\Psi_{\hat{x}}\,(B_{1u}) = \hat{x}\left(\psi_{26}^t - \psi_{27}^t - \psi_1^t + \psi_4^t\right) + (-\hat{y})\left(-\psi_{37}^t + \psi_{40}^t - \psi_{14}^t + \psi_{15}^t\right) = -\Psi_{\hat{x}}\,(B_{1u})\\
&h_{27}\Psi_{\hat{x}}\,(B_{1u}) = \hat{x}\left(\psi_{27}^t - \psi_{26}^t - \psi_1^t + \psi_4^t\right) + (-\hat{y})\left(-\psi_{40}^t + \psi_{37}^t - \psi_{15}^t + \psi_{14}^t\right) = -\Psi_{\hat{x}}\,(B_{1u})\\
&Ih_4\Psi_{\hat{x}}\,(B_{1u}) = I\,(-\hat{x})\left(\psi_4^t - \psi_1^t - \psi_{27}^t + \psi_{26}^t\right) + I\,(-\hat{y})\left(-\psi_{15}^t + \psi_{14}^t - \psi_{40}^t + \psi_{37}^t\right) = -\Psi_{\hat{x}}\,(B_{1u})\\
&h_{26}\Psi_{\hat{y}}\,(B_{1u}) = (-\hat{y})\left(\psi_{26}^t - \psi_{27}^t + \psi_1^t - \psi_4^t\right) + \hat{x}\left(\psi_{37}^t - \psi_{40}^t - \psi_{14}^t + \psi_{15}^t\right) = -\Psi_{\hat{y}}\,(B_{1u})\\
&h_{27}\Psi_{\hat{y}}\,(B_{1u}) = \hat{y}\left(\psi_{27}^t - \psi_{26}^t + \psi_4^t - \psi_1^t\right) + (-\hat{x})\left(\psi_{40}^t - \psi_{37}^t - \psi_{15}^t + \psi_{14}^t\right) = -\Psi_{\hat{y}}\,(B_{1u})\\
&Ih_4\Psi_{\hat{y}}\,(B_{1u}) = I\,(-\hat{y})\left(\psi_4^t - \psi_1^t + \psi_{27}^t - \psi_{26}^t\right) + I\,(-\hat{x})\left(\psi_{15}^t - \psi_{14}^t - \psi_{40}^t + \psi_{37}^t\right) = -\Psi_{\hat{y}}\,(B_{1u})\\
&h_{26}\Psi_{\hat{z}}\,(B_{1u}) = (-\hat{z})\left(\psi_{26}^t + \psi_{27}^t + \psi_1^t + \psi_4^t - \psi_{37}^t - \psi_{40}^t - \psi_{14}^t - \psi_{15}^t\right) = -\Psi_{\hat{z}}\,(B_{1u})\\
&h_{27}\Psi_{\hat{z}}\,(B_{1u}) = (-\hat{z})\left(\psi_{27}^t + \psi_{26}^t + \psi_4^t + \psi_1^t - \psi_{40}^t - \psi_{37}^t - \psi_{15}^t - \psi_{14}^t\right) = -\Psi_{\hat{z}}\,(B_{1u})\\
&Ih_4\Psi_{\hat{z}}\,(B_{1u}) = I\hat{z}\left(\psi_4^t + \psi_1^t + \psi_{27}^t + \psi_{26}^t - \psi_{15}^t - \psi_{14}^t - \psi_{40}^t - \psi_{37}^t\right) = -\Psi_{\hat{z}}\,(B_{1u})
\end{aligned}
$$

Finally, we obtain nodal structure for B_{2u}:

$$h_{26}\Psi_{\hat{x}}(B_{2u}) = \hat{x}\left(\psi_{26}^t - \psi_{27}^t + \psi_1^t - \psi_4^t\right) + (-\hat{y})\left(-\psi_{37}^t + \psi_{40}^t + \psi_{14}^t - \psi_{15}^t\right) = \Psi_{\hat{x}}(B_{2u})$$
$$h_{27}\Psi_{\hat{x}}(B_{2u}) = (-\hat{x})\left(\psi_{27}^t - \psi_{26}^t + \psi_4^t - \psi_1^t\right) + \hat{y}\left(-\psi_{40}^t + \psi_{37}^t + \psi_{15}^t - \psi_{14}^t\right) = \Psi_{\hat{x}}(B_{2u})$$
$$Ih_4\Psi_{\hat{x}}(B_{2u}) = I(-\hat{x})\left(\psi_4^t - \psi_1^t + \psi_{27}^t - \psi_{26}^t\right) + I(-\hat{y})\left(-\psi_{15}^t + \psi_{14}^t + \psi_{40}^t - \psi_{37}^t\right) = -\Psi_{\hat{x}}(B_{2u})$$
$$h_{26}\Psi_{\hat{y}}(B_{2u}) = (-\hat{y})\left(\psi_{26}^t - \psi_{27}^t - \psi_1^t + \psi_4^t\right) + \hat{x}\left(\psi_{37}^t - \psi_{40}^t + \psi_{14}^t - \psi_{13}^t\right) = \Psi_{\hat{y}}(B_{2u})$$
$$h_{27}\Psi_{\hat{y}}(B_{2u}) = \hat{y}\left(\psi_{27}^t - \psi_{26}^t - \psi_4^t + \psi_1^t\right) + (-\hat{x})\left(\psi_{40}^t - \psi_{37}^t + \psi_{15}^t - \psi_{14}^t\right) = \Psi_{\hat{y}}(B_{2u})$$
$$Ih_4\Psi_{\hat{y}}(B_{2u}) = I(-\hat{y})\left(\psi_4^t - \psi_1^t - \psi_{27}^t + \psi_{26}^t\right) + I(-\hat{x})\left(\psi_{15}^t - \psi_{14}^t + \psi_{40}^t - \psi_{37}^t\right) = -\Psi_{\hat{y}}(B_{2u})$$
$$h_{26}\Psi_{\hat{z}}(B_{2u}) = (-\hat{z})\left(\psi_{26}^t + \psi_{27}^t - \psi_1^t - \psi_4^t - \psi_{37}^t - \psi_{40}^t + \psi_{14}^t + \psi_{15}^t\right) = \Psi_{\hat{z}}(B_{2u})$$
$$h_{27}\Psi_{\hat{z}}(B_{2u}) = (-\hat{z})\left(\psi_{27}^t + \psi_{26}^t - \psi_1^t - \psi_4^t - \psi_{40}^t - \psi_{37}^t + \psi_{15}^t + \psi_{14}^t\right) = \Psi_{\hat{z}}(B_{2u})$$
$$Ih_4\Psi_{\hat{z}}(B_{2u}) = I\hat{z}\left(\psi_4^t + \psi_1^t - \psi_{27}^t - \psi_{26}^t - \psi_{15}^t - \psi_{14}^t + \psi_{40}^t + \psi_{37}^t\right) = -\Psi_{\hat{z}}(B_{2u})$$

The nodal structure of these functions on coordinate planes (100), (010) and (001) is presented in Table 4.18.

It is seen in Fig. 4.10 that at diagonal plane the function with different spin projections $\hat{x}$ and $\hat{y}$ meet and there should be nodes for all IRs. However, this assertion is not general. If the triplet spin is quantized in k_z-direction the triplet spin projections $M_s = \pm 1$ are expressed as complex linear combinations $\hat{x} \pm i\hat{y}$. The extension of the technique in this case is considered in Sect. 4.12. As to the functions with $\hat{z}$ triplet spin component, their nodal structure in vertical diagonal planes can be easily obtained by the same technique as follows:

$$h_{37}\Psi_{\hat{z}}(A_{1u}) = (-\hat{z})\left(\psi_{37}^t + \psi_{40}^t + \psi_{15}^t + \psi_{14}^t + \psi_{27}^t + \psi_{26}^t + \psi_1^t + \psi_4^t\right) = -\psi_{\hat{z}}(A_{1u})$$
$$h_{37}\Psi_{\hat{z}}(A_{2u}) = (-\hat{z})\left(\psi_{37}^t + \psi_{40}^t - \psi_{15}^t - \psi_{14}^t + \psi_{27}^t + \psi_{26}^t - \psi_1^t - \psi_4^t\right) = \psi_{\hat{z}}(A_{2u})$$
$$h_{37}\Psi_{\hat{z}}(B_{1u}) = (-\hat{z})\left(\psi_{37}^t + \psi_{40}^t + \psi_{15}^t + \psi_{14}^t - \psi_{27}^t - \psi_{26}^t - \psi_1^t - \psi_4^t\right) = \psi_{\hat{z}}(B_{1u})$$
$$h_{37}\Psi_{\hat{z}}(B_{2u}) = (-\hat{z})\left(\psi_{37}^t + \psi_{40}^t - \psi_{15}^t - \psi_{14}^t - \psi_{27}^t - \psi_{26}^t + \psi_1^t + \psi_4^t\right) = -\psi_{\hat{z}}(B_{2u})$$

Two real ESP functions $\hat{x}$ and $\hat{y}$ belong to real IR E_g and two complex ESP functions $\hat{x} \pm i\hat{y}$ belong to complex IR E_g and their transformation properties similar to those of spatial parts belonging to the same IRs. It is the subject of the next two sections.

4.9 Additional Quantum Numbers for Real Two-Dimensional IRs

According to Mackey theorem at general $k-$point in a BZ, the expansion of the symmetrized (antisymmetrized) square includes IR E_g (E_u) of D_{4h} group two times. It follows from the above analysis that on vertical planes one two-dimensional IR $E_g(E_u)$ is allowed, and the other IR $E_g(E_u)$ is prohibited. Hence, the nodes in the vertical planes are not "symmetrically protected". However, these nodal planes can be classified group-theoretically in terms of IRs of intermediate subgroup of index 2, namely D_{2h} or C_{4h} group. For k a general point of a BZ the group in Mackey theorem $\tilde{M} = \mathrm{C}_i$. Consider a two-step induction of odd IR A_u from C_i to D_{4h}. In the first step, we induce A_u from C_i into C_{2h} and obtain

$$A_u \uparrow C_{2h} = A_u + B_u \tag{4.9.1}$$

Let C_{2h} be the symmetry group of plane (010). Then further inductions from C_{2h} into D_{4h} result in

$$A_u \uparrow D_{4h} = A_{1u} + B_{1u} + E_u \tag{4.9.2}$$

$$B_u \uparrow D_{4h} = A_{2u} + B_{2u} + E_u \tag{4.9.3a}$$

The r.h.s. of formula (4.9.3a) coincides with the set of pair functions on the plane (010) in Table 4.12. It means that IRs A_{2u}, B_{2u} and E_u are nodeless on this plane. Since the spatial part of triplet pair belongs to B_u on the plane (see Sect. 4.3), we can conclude that the result of induction of A_u into D_{4h}, i.e. IRs $A_{1u} + B_{1u} + E_u$ in r.h.s. of formula (4.9.2) are nodal on the plane (010). The basis functions of IRs in r.h.s. of (4.9.1) near the plane are written as

$$\Psi(A_u) = \left(\varphi_{k_1}\varphi_{Ik_1} - \varphi_{Ik_1}\varphi_{k_1}\right) - \left(\varphi_{\sigma_y k_1}\varphi_{C_{2y}k_1} - \varphi_{C_{2y}k_1}\varphi_{\sigma_y k_1}\right) \tag{4.9.4}$$

$$\Psi(B_u) = \left(\varphi_{k_1}\varphi_{Ik_1} - \varphi_{Ik_1}\varphi_{k_1}\right) + \left(\varphi_{\sigma_y k_1}\varphi_{C_{2y}k_1} - \varphi_{C_{2y}k_1}\varphi_{\sigma_y k_1}\right) \tag{4.9.5}$$

When k_1 approaches symmetry plane (010) the wavevector $\sigma_y k_1$ approaches k_1 and $C_{2y}k_1$ approaches Ik_1. In this case two terms of $\Psi(A_u)$ cancel and two terms of $\Psi(B_u)$ merge resulting in nodal and nodeless structure on this plane respectively.

Going over to two-dimensional IR we see that one E_u (in (4.9.2)) corresponds to A_u near the plane and another E_u (in (4.9.3a)) corresponds to B_u near the plane and on the plane (010). This means that there are two basis sets of E_u symmetry, nodal and nodeless on the plane, which can by labeled by the additional quantum number, i.e. the label of IR of an intermediate group C_{2h}. However, in the case of D_{4h} group, it is more convenient to consider a larger intermediate group D_{2h}. When inducing from C_{2h} into intermediate subgroup D_{2h}, we obtain

$$A_{u(C_{2h})} \uparrow D_{2h} = A_u + B_{2u} \tag{4.9.6}$$

$$B_{u(C_{2h})} \uparrow D_{2h} = B_{1u} + B_{3u} \tag{4.9.7}$$

Further induction into D_{4h} results in

$$B_{2u(D_{2h})} \uparrow D_{4h} = E_u \tag{4.9.8}$$

$$B_{3u(D_{2h})} \uparrow D_{4h} = E_u \tag{4.9.9}$$

It follows from these relations that the induction of basis sets of IRs B_{2u} and B_{3u} of D_{2h} into D_{4h} results in irreducible basis sets of E_u symmetry. This means that total basis set on group D_{4h} can be obtained acting by the left coset representative on the basis set on group D_{2h}. The E_u nodal (nodeless) basis stems from B_{2u} (B_{3u}) of D_{2h} which has a character -1 ($+1$) for the reflection h_{27} in (010) plane.

Similar analysis may be performed for even IRs. When inducing from C_{2h} into intermediate subgroup D_{2h} we obtain

$$A_{g(C_{2h})} \uparrow D_{2h} = A_g + B_{2g} \tag{4.9.10}$$

$$B_{g(C_{2h})} \uparrow D_{2h} = B_{1g} + B_{3g} \tag{4.9.11}$$

Further induction into D_{4h} results in

$$B_{2g(D_{2h})} \uparrow D_{4h} = E_g \tag{4.9.12}$$

$$B_{3g(D_{2h})} \uparrow D_{4h} = E_g \tag{4.9.13a}$$

Two first equations correspond to nodeless states and the last two equations correspond to nodal states. Thus the E_g basis set nodal (nodeless) on plane (010) basis stems from B_{3g} (B_{2g}) of D_{2h} which has a character -1 ($+1$) for the reflection h_{27} in (010) plane.

The basis sets of triplet pairs on group D_{2h} are easily obtained by projection operators and they are written as

$$\Psi_1(B_{3u}) = \psi_1^t - \psi_4^t - \psi_{26}^t + \psi_{27}^t \tag{4.9.14a}$$

$$\Psi_1(B_{2u}) = \psi_1^t - \psi_4^t + \psi_{26}^t - \psi_{27}^t \tag{4.9.15a}$$

Acting by reflection h_{27} on these sets, we make sure that the first basis set is nodeless on the plane (010) and the second is nodal. When choosing the reflection h_{37} as the left coset representative in the induction from D_{2h} into D_{4h} we obtain basis functions of second rows of induced representations (4.9.9) and (4.9.8):

$$\Psi_2(B_{3u}) = h_{37}\Psi_1(B_{3u}) = -\psi_{37}^t + \psi_{40}^t + \psi_{15}^t - \psi_{14}^t \tag{4.9.16a}$$

$$\Psi_2(B_{2u}) = h_{37}\Psi_1(B_{2u}) = -\psi_{37}^t + \psi_{40}^t - \psi_{15}^t + \psi_{14}^t \tag{4.9.17a}$$

The minus sign in these formulas corresponds to minus sign in matrix $E_u(h_{37})$ in Table 4.19. However, as it will be shown below, one can also take a plus sign for this matrix and the total number of real E_u basis sets equals to four. It is seen in Fig. 4.7 that when k_1 approaches k_x-axis (plane (010) in three dimensions) the functions of the second row of E_u meet at k_y (plane (100) in three dimensions). The nodal structure at the plane (100) may be obtained by action of h_{26} (reflection in plane (100):

Table 4.19 Matrices of IR E_u of the elements of D_{4h} on its subgroup C_{4v} in real form

$E=h_1$	$C_{2z}=h_4$	$\sigma_x=h_{26}$	$\sigma_y=h_{27}$
$\begin{pmatrix}1&0\\0&1\end{pmatrix}$	$\begin{pmatrix}-1&0\\0&-1\end{pmatrix}$	$\begin{pmatrix}-1&0\\0&1\end{pmatrix}$	$\begin{pmatrix}1&0\\0&-1\end{pmatrix}$
$\sigma_b=h_{37}$	$\sigma_a=h_{40}$	$C_{4z}=h_{14}$	$C_{4z}^3=h_{15}$
$\begin{pmatrix}0&-1\\-1&0\end{pmatrix}$	$\begin{pmatrix}0&1\\1&0\end{pmatrix}$	$\begin{pmatrix}0&1\\-1&0\end{pmatrix}$	$\begin{pmatrix}0&-1\\1&0\end{pmatrix}$

$$h_{26}\Psi_2(B_{3u}) = h_{26}(-\psi_{37}^t + \psi_{40}^t + \psi_{15}^t - \psi_{14}^t) = -\psi_{14}^t + \psi_{15}^t + \psi_{40}^t - \psi_{37}^t = \Psi_2(B_{3u})$$

$$h_{26}\Psi_2(B_{2u}) = h_{26}(-\psi_{37}^t + \psi_{40}^t - \psi_{15}^t + \psi_{14}^t) = -\psi_{14}^t + \psi_{15}^t - \psi_{40}^t + \psi_{37}^t = -\Psi_2(B_{2u})$$

The function $\Psi_2(B_{2u})$ changed the sign, but the function $\Psi_2(B_{3u})$ is invariant under the action of h_{26}. This implies that the function $\Psi_2(B_{2u})$ is nodal and the function $\Psi_2(B_{3u})$ is nodeless at the plane (100). Thus we proved that in D_{4h} group the nodal (nodeless) function of E_u symmetry on one vertical plane will be nodal (nodeless) on the other equivalent vertical plane.

However, there is a more straightforward way to obtain all possible basis functions of real two-dimensional IRs E_u and E_g. It is immediately verified that a direct product of E_u (or E_g) by any even one-dimensional IR results in also E_u (or E_g). Multiplication by A_{2g} changes the signs for rotations in all horizontal axes and hence changes the signs for reflections in all vertical planes, resulting in reversing of nodal structure in vertical planes. Similarly multiplication by B_{1g} or B_{2g} changes the signs of half of rotations about horizontal axes and half of reflections in vertical planes Thus in order to obtain all possible nodal structures of two-electron states belonging to IR E_u it is sufficient to project basis set shown in Fig. 4.7 on the standard form of E_u, (or E_g) [60], and then to multiply the result by all one-dimensional even IRs of D_{4h}.

Matrices of IR E_u of D_{4h} in real (coordinate) form can be easily constructed by considering transformations of coordinate basis (x, y), where the first line corresponds to x and the second line corresponds to y. These matrices are presented in Table 4.19.

Making use of these matrices and projection operators (4.8.5) one obtains spatial parts of triplet pairs denoted as $\Psi_{1,2}\,(E_u)$ presented in Table 4.20. Since direct products of E_u by any of one-dimensional even IRs results in also E_u, one can use these IRs to construct two-electron basis sets, whose nodal structures differ from that of standard E_u. These basis sets are denoted in Table 4.20 by superscripts A_{2g}, B_{1u} and B_{2g}.

The nodal structure on the symmetry planes can be estimated as follows. Since the pair function on the plane is invariant under the plane reflection, the nodeless linear combination remains unchanged under the action of the plane reflection. If the reflection changes the sign of the function, then this function vanishes on the plane, i.e. it is nodal on this plane.

Consider, for example, function $E_u^{A_{2g}}$. The action of reflection in the (010) plane on the function of the first line of E_u does changes the sign of the function:

Table 4.20 Spatial parts of triplet pairs for the IR E_u of group D_{4h} in real representation and their nodal planes (last two columns). The element h_{37} is chosen as a left coset representative in the induction. Basis functions are given by formula (4.7.1) and the basis set is shown Fig. 4.7

IR	Function	Nodes in planes	
		Type 1[1]	Type 2[2]
$\Psi_1(E_u)$	$\psi_1^t - \psi_4^t - \psi_{26}^t + \psi_{27}^t$	No	$(\bar{1}10)$
$\Psi_2(E_u)$	$-\psi_{37}^t + \psi_{40}^t + \psi_{15}^t - \psi_{14}^t$	No	
$\Psi_1\left(E_u^{A_{2g}}\right)$	$\psi_1^t - \psi_4^t + \psi_{26}^t - \psi_{27}^t$	(010)	(110)
$\Psi_2\left(E_u^{A_{2g}}\right)$	$\psi_{37}^t - \psi_{40}^t + \psi_{15}^t - \psi_{14}^t$	(100)	
$\Psi_1\left(E_u^{B_{1g}}\right)$	$\psi_1^t - \psi_4^t - \psi_{26}^t + \psi_{27}^t$	No	(110)
$\Psi_2\left(E_u^{B_{1g}}\right)$	$\psi_{37}^t - \psi_{40}^t - \psi_{15}^t + \psi_{14}^t$	No	
$\Psi_1\left(E_u^{B_{2g}}\right)$	$\psi_1^t - \psi_4^t + \psi_{26}^t - \psi_{27}^t$	(010)	$(\bar{1}10)$
$\Psi_2\left(E_u^{B_{2g}}\right)$	$-\psi_{37}^t + \psi_{40}^t - \psi_{15}^t + \psi_{14}^t$	(100)	

[1] Nodes due to cancellation of basis functions of one row
[2] Nodes due to cancellation of basis functions of two rows

$$h_{27}\Psi_1\left(E_u^{A_{2g}}\right) = h_{27}\left(\psi_1^t - \psi_4^t + \psi_{26}^t - \psi_{27}^t\right) = \psi_{27}^t - \psi_{26}^t + \psi_4^t - \psi_1^t \quad (4.9.18)$$

Since the function changed the sign, we conclude that the function of the first row has nodes in this plane. It is seen in Fig. 4.7 that when k_1 approaches the plane (010) of the BZ, the functions of the second row meet around the (100) plane. When acting on the second row function by $h_{26} = \sigma_x$ element, we find that this function also changes the sign:

$$h_{26}\Psi_2\left(E_u^{A_{2g}}\right) = h_{26}\left(\psi_{37}^t - \psi_{40}^t + \psi_{15}^t - \psi_{14}^t\right) = \psi_{14} - \psi_{15} + \psi_{40} - \psi_{37} \quad (4.9.19)$$

Hence, it follows that the function of the second row is nodal in the plane (100).

The zeros of the basis functions of individual rows of a two-dimensional IR will be called zeros of the first type. In the case of two-dimensional IRs, there is also the possibility that the function of a pair vanishes if the functions of two rows are cancelled. Such zeros will be called zeros of the second type. As can be seen in Fig. 4.7, this can happen on the diagonal planes (110) and (−110). Acting with the element h_{37} on the sum of two rows of $\Psi\left(E_u^{A_{2g}}\right)$, we obtain

$$h_{37}\left(\psi_1^t - \psi_4^t + \psi_{26}^t - \psi_{27}^t + \psi_{37}^t - \psi_{40}^t + \psi_{15}^t - \psi_{14}^t\right) = \psi_{37}^t - \psi_{40}^t + \psi_{15}^t - \psi_{14}^t + \psi_1^t - \psi_4^t + \psi_{26}^t - \psi_{27}^t \quad (4.9.20)$$

As a result of reflection in the plane (−110), the sign of the function was not changed so there are no nodes of the second type on the plane (−110). To check the possibility of the existence of zeros on the second diagonal plane (110), we act on the sum of the two rows of $\Psi\left(E_u^{A_{2g}}\right)$ by the operator h_{40}:

$$h_{40}\left(\psi_1^t - \psi_4^t + \psi_{26}^t - \psi_{27}^t + \psi_{37}^t - \psi_{40}^t + \psi_{15}^t - \psi_{14}^t\right) = \psi_{40}^t - \psi_{37}^t + \psi_{14}^t - \psi_{15}^t + \psi_4^t - \psi_1^t + \psi_{27}^t - \psi_{26}^t \quad (4.9.21)$$

As a result of reflection in the plane (110), the sign of the function was changed, hence it follows that the function $\Psi\left(E_u^{A_{2g}}\right)$ has nodes on the (110) plane.

When checking the presence of zeros on the basal plane of the function $\Psi\left(E_u^{A_{2g}}\right)$, it is necessary to act with the group element $h_{28} = Ih_4$ (reflection in plane (001)) on both basis functions separately:

$$Ih_4\Psi_1\left(E_u^{A_{2g}}\right) = Ih_4\left(\psi_1^t - \psi_4^t + \psi_{26}^t - \psi_{27}^t\right) = I\left(\psi_4^t - \psi_1^t + \psi_{27}^t - \psi_{26}^t\right) = -\psi_4^t + \psi_1^t - \psi_{27}^t + \psi_{26}^t \quad (4.9.22)$$

$$Ih_4\Psi_2\left(E_u^{A_{2g}}\right) = Ih_4\left(\psi_{37}^t - \psi_{40}^t + \psi_{15}^t - \psi_{14}^t\right) = I\left(\psi_{40}^t - \psi_{37}^t + \psi_{14}^t - \psi_{15}^t\right) = -\psi_{40}^t + \psi_{37}^t - \psi_{14}^t + \psi_{15}^t \quad (4.9.23)$$

Thus, the basis functions remained unchanged and there are no zeros on the basal plane (001).

In basis sets of E_u type presented in Table 4.20 together with their nodal structure, it is assumed that the wavevector k_1 of the first function ψ_1 can take any value in the basis domain of the BZ, i.e. $0 \le \theta \le \pi/4$, $0 \le k_z \le b_3/2$. When k_1 runs over the sector $0 \le \theta \le \pi/4$, the functions of the first line run over the white area, and the functions of the second line run over the shadowed area in Fig. 4.11. Nodes of both types are shown in Fig. 4.11 by bold red lines. The basis sets $E_u^{A_{2g}}$ and $E_u^{B_{2g}}$ have nodes of the first type in two vertical coordinate planes, while the basis sets E_u and $E_u^{B_{1g}}$ do not have such nodes. However all basis sets have nodes in one of the diagonal planes.

Singlet functions are obtained by projecting on the IR E_g, so the coefficients in linear combinations are obtained by changing the sign for elements with inversion, i.e. for h_i, if $i \ge 25$. With such a replacement, the nodal structure is reversed. For example for the first function we have

$$h_{27}\Psi_1\left(E_g\right) = h_{27}\left(\psi_1^s - \psi_4^s + \psi_{26}^s - \psi_{27}^s\right) = \psi_{27}^s - \psi_{26}^s + \psi_4^s - \psi_1^s \quad (4.9.24)$$

$$h_{26}\Psi_2\left(E_g\right) = h_{26}\left(-\psi_{37}^s + \psi_{40}^s - \psi_{14}^s + \psi_{15}^s\right) = \psi_{14}^s - \psi_{15}^s + \psi_{37}^s - \psi_{40}^s \quad (4.9.25)$$

The reflections in coordinate planes changed the signs of two basis functions, thus the function $\Psi\left(E_g\right)$ is nodal in vertical coordinate plane. When acting by reflection

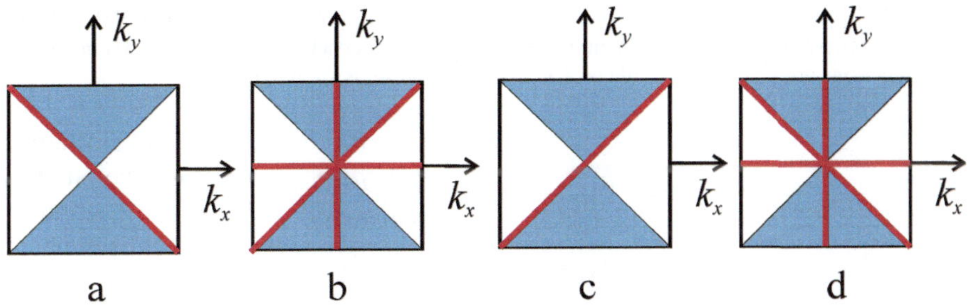

Fig. 4.11 Nodes (red lines) of odd basis functions in k-space. White area—function of the first line, shadowed area—function of the second line (see Table 4.20), **a** E_u **b** $E_u^{A_{2g}}$ **c** $E_u^{B_{1g}}$ **d** $E_u^{B_{2g}}$

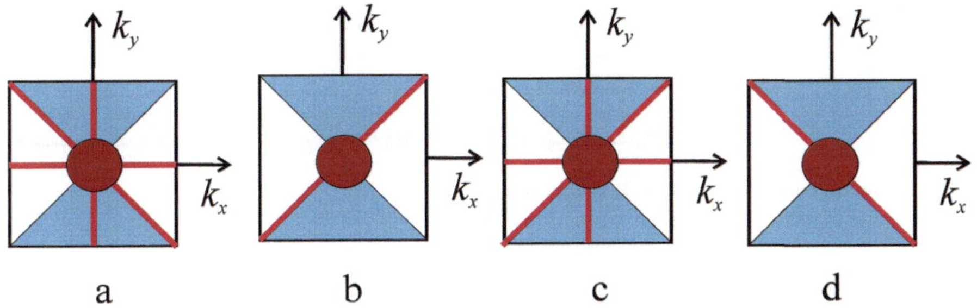

Fig. 4.12 Nodes (red lines) of odd basis functions in k-space in plane normal to k_z. Red circles denote nodes in basal plane $k_z = 0$. White area—function of the first line, shadowed area—function of the second line (see Table 4.21), **a** E_g **b** $E_g^{A_{2g}}$ **c** $E_g^{B_{1g}}$ **d** $E_g^{B_{2g}}$

in the basal plane $h_{28} = Ih_4$ on singlet functions, and taking into account that they are even, we obtain

$$Ih_4\Psi_1\left(E_g\right) = Ih_4\left(\psi_1^s - \psi_4^s + \psi_{26}^s - \psi_{27}^s\right) = \psi_4^s - \psi_1^s + \psi_{27}^s - \psi_{26}^s \quad (4.9.26)$$

$$Ih_4\Psi_2\left(E_g\right)\left(\psi_{37}^s - \psi_{40}^s + \psi_{15}^s - \psi_{14}^s\right) = \psi_{40}^s - \psi_{37}^s + \psi_{14}^s - \psi_{15}^s \quad (4.9.27)$$

The sign was changed for both basis functions $\Psi\left(E_g\right)$ and this function is nodal in the basal plane.

As it seen in Table 4.21, where singlet basis sets for IR E_g are presented, the nodal structure on vertical planes is similar to that in a triplet case (see Fig. 4.10): the nodal structures in vertical planes of E_g, $E_g^{A_{2g}}$, $E_g^{B_{1g}}$ and $E_g^{B_{2g}}$ are the same as that of $E_u^{A_{2g}}$, E_u, $E_u^{B_{2g}}$ and $E_u^{B_{1g}}$, respectively. An important difference is that all singlet bases of the E_g symmetry have nodes in the (001) plane, while the triplet pairs of E_u symmetry are nodeless in the basal plane.

Possible nodal structures for even two-dimensional IR E_g are shown in Fig. 4.12. When comparing with the results for odd IR E_u presented in Fig. 4.11 we see that

Table 4.21 Spatial parts of singlet pairs for real representation E_g of D_{4h} group and their nodal planes (last two columns). The element h_{37} is chosen as a left coset representative in the induction. Basis functions are given by formula (4.7.1) and the basis set is shown in Fig. 4.7

		Nodes	
IR	Function	Type 1[1)]	Type 2[2)]
$\Psi_1(E_g)$	$\psi_1^s - \psi_4^s + \psi_{26}^s - \psi_{27}^s$	(010), (001)	(110)
$\Psi_2(E_g)$	$\psi_{37}^s - \psi_{40}^s + \psi_{15}^s - \psi_{14}^s$	(100), (001)	
$\Psi_1\left(E_g^{A_{2g}}\right)$	$\psi_1^s - \psi_4^s - \psi_{26}^s + \psi_{27}^s$	(001)	$(\bar{1}10)$
$\Psi_2\left(E_g^{A_{2g}}\right)$	$-\psi_{37}^s + \psi_{40}^s + \psi_{15}^s - \psi_{14}^s$	(001)	
$\Psi_1\left(E_g^{B_{1g}}\right)$	$\psi_1^s - \psi_4^s + \psi_{26}^s - \psi_{27}^s$	(010), (001)	$(\bar{1}10)$
$\Psi_2\left(E_g^{B_{1g}}\right)$	$-\psi_{37}^s + \psi_{40}^s - \psi_{15}^s + \psi_{14}^s$	(100), (001)	
$\Psi_1\left(E_g^{B_{2g}}\right)$	$\psi_1^s - \psi_4^s - \psi_{26}^s + \psi_{27}^s$	(001)	(110)
$\Psi_2\left(E_g^{B_{2g}}\right)$	$\psi_{37}^s - \psi_{40}^s - \psi_{15}^s + \psi_{14}^s$	(001)	

1) Nodes due to cancellation of basis functions of one row
2) Nodes due to cancellation of basis functions of two rows

nodal structures in vertical coordinate planes are very similar, i.e. there are functions nodal or nodeless in both vertical coordinate planes, but in all cases there is one nodal diagonal plane. The essential difference exists in the basal plane: E_u is always nodeless but E_g is always nodal in the basal plane.

4.10 Angular Momentum of SOP

In axial point groups D_{nh} the characters of one-dimensional IRs $D^k(\bar{\theta})$ for the rotations about vertical axis may be related to quasi-angular momentum $\bar{m}$ by the relation:

$$D^k(\bar{\theta}) = \exp(i\bar{m}\bar{\theta}) \tag{4.10.1}$$

However, in the case of two-dimensional IRs in real form, considered in the previous section, this relation is not obvious. To find a convenient form of two-dimensional IRs, we will use intermediate subgroups C_{nh} of index 2, whose IRs are one-dimensional and are directly connected with angular momentum $\bar{m}$. In this section, we take group D_{4h} as an example. It is seen from Table 4.22, where IRs of C_{4h} group are presented, that one-dimensional IRs $E'_{1g(u)}$ and $E'_{2g(u)}$ correspond to $\bar{m} = 1$ and $\bar{m} = -1$, respectively, $B_{g(u)}$ correspond to $\bar{m} = \pm 2$ and $A_{g(u)}$ can be associated with both $\bar{m} = 0$ and $\bar{m} = \pm 4$. IRs of the entire group, obtained by the induction from C_{4h} into D_{4h}, presented in the last column, retain the same connection with angular momentum. It is remarkable that the induction of complex IRs results in irreducible complex representations $\bar{E}_{g(u)}$. Note that since all characters of $\tilde{E}_{g(u)}$ are real, it is equivalent to complex conjugate representation $\tilde{E}^*_{g(u)}$.

Consider the construction of complex induced representation $\tilde{E}_g$ of group D_{4h} on the basis of complex IR E'_{1g} of its subgroup C_{4h}. We choose the reflection $\sigma_y = h_{27}$ in plane (010) as a left coset representative. For the elements of subgroup C_4 in the first line of the matrix $\tilde{E}_{g(u)}$, on the main diagonal there is IR E'_{1g} of C_{4h} group. To obtain the elements of the second line of this IR, we find conjugated elements $h_{27}^{-1}h_{14}h_{27} = h_{15}$, $h_{27}^{-1}h_{15}h_{27} = h_{14}$ and $h_{27}^{-1}h_4h_{27} = h_4$. Hence, it follows that in the second row on the main diagonal of the matrix $\tilde{E}_{g(u)}$ for the element h_4 there is the IR $E'_{1g}(h_4)$, and for the elements h_{14} and h_{15} there are permuted elements of the first row. For the matrix of the left coset element h_{27}, we have $h_1h_{27}h_{27} = h_1$ and $h_{27}^{-1}h_{27}h_1 = h_1$. Hence, it follows that off-diagonal matrix elements of $\tilde{E}_{g(u)}(h_{27})$ are equal to 1. These matrices are presented in Table 4.23. It is seen in Table 4.23 that for the rotations C_4 matrices $\tilde{E}_{g(u)}$ are diagonal, the first row corresponds to $\bar{m} = +1$ and the second row corresponds to $\bar{m} = -1$. When multiplying the matrices for rotation elements by the matrix $\tilde{E}_{g(u)}(h_{27})$ we obtain all matrices for the elements of subgroup C_{4v}. Note that so far all matrices $\tilde{E}_g$ and $\tilde{E}_u$ were identical. The representations $\tilde{E}_g$ and $\tilde{E}_u$ are extended on the left coset IC_{4v} with signs + and −, respectively. When multiplying the matrices $\tilde{E}_g$ and $\tilde{E}_u$ by IRs A_{2g}, B_{1g} and B_{2u} we obtain other chiral forms, differing by nodal structure and phase winding direction. Namely, multiplication by A_{2g} reverses the nodal structure, multiplication by B_{1g} changes the phase winding direction of the rows and multiplication by B_{2u} changes both nodal structure and phase winding direction. The labels of this IRs are used as the superscripts. Note that we used the same matrix of left coset representative h_{27} for even and odd IRs and in our notations $\tilde{E}_u = \tilde{E}_g \times A_{2u}$. This relation differs from standard one: $E_u = E_g \times A_{1u}$. However, since all possible forms of complex two-dimensional IRs are considered anyway, this relation means only their renaming. In this notation, vertical nodes and deeps of $\tilde{E}_g$ and $\tilde{E}_u$ with the same superscripts are identical. The matrices are presented in Table 4.23.

The pairs' functions of $\tilde{E}_g$ and $\tilde{E}_u$ symmetry constructed from basis set shown in Fig. 4.7 making use of projection operators (4.7.2) are presented in Table 4.24.

Table 4.22 IRs of the group C_{4h} on the subgroup C_4 and decomposition of the corresponding induced representations. Tilde is used to distinguish complex two-dimensional IRs from real ones presented in Table 4.19

D^k	E	C_{4z}	C_{2z}	C^3_{4z}	$\bar{m}$	$D^k \uparrow D_{4h}$
$A_{g(u)}$	1	1	1	1	0, ±4	$A_{1g(u)} + A_{2g(u)}$
$B_{g(u)}$	1	−1	1	−1	±2	$B_{1g(u)} + B_{2g(u)}$
$E'_{1g(u)}$	1	i	−1	$-i$	1	$\tilde{E}_{g(u)}$
$E'_{2g(u)}$	1	$-i$	−1	i	1	$\tilde{E}^*_{g(u)}$

Table 4.23 Selected matrices of IRs $\tilde{E}_{g(u)}$ of D_{4h} group in complex form. To obtain the matrices on the subgroup C_{4v} it is sufficient to multiply all matrices for pure rotations by the matrix of the left coset representative h_{27}. Note that on the subgroup C_{4v} matrices of $\tilde{E}_g$ and $\tilde{E}_u$ are chosen to be the same. The matrices $\tilde{E}_g$ and $\tilde{E}_u$ may be easily extended on D_{4h} with multiplying by inversion I with the signs + or −, respectively

h_1	h_{14}	h_4	h_{15}	h_{27}
$\begin{pmatrix} 1 & 0 \\ 0 & 1 \end{pmatrix}$	$\begin{pmatrix} i & 0 \\ 0 & -i \end{pmatrix}$	$\begin{pmatrix} -1 & 0 \\ 0 & -1 \end{pmatrix}$	$\begin{pmatrix} -i & 0 \\ 0 & i \end{pmatrix}$	$\begin{pmatrix} 0 & 1 \\ 1 & 0 \end{pmatrix}$

Table 4.24 Basis functions of two-dimensional IRs $\tilde{E}_{g(u)}$ in complex form. Basis functions correspond to Fig. 4.7

		Nodes		Dips
IR	Function	Type [1)]	Type [2)]	Type [2)]
$\Psi_1\left(\tilde{E}_{g(u)}\right)$	$\psi_1^{s(t)} + i\psi_{14}^{s(t)} - \psi_4^{s(t)} - i\psi_{15}^{s(t)}$	$(001)(s)$	(010)	$(110), (\bar{1}10)$
$\Psi_2\left(\tilde{E}_{g(u)}\right)$	$\psi_{27}^{s(t)} - i\psi_{37}^{s(t)} - \psi_{26}^{s(t)} + i\psi_{40}^{s(t)}$	$(001)(s)$		
$\Psi_1\left(\tilde{E}_{g(u)}^{A_{2g}}\right)$	$\psi_1^{s(t)} + i\psi_{14}^{s(t)} - \psi_4^{s(t)} - i\psi_{15}^{s(t)}$	$(001)(s)$	(100)	$(110), (\bar{1}10)$
$\Psi_2\left(\tilde{E}_{g(u)}^{A_{2g}}\right)$	$-\psi_{27}^{s(t)} + i\psi_{37}^{s(t)} + \psi_{26}^{s(t)} - i\psi_{40}^{s(t)}$	$(001)(s)$		
$\Psi_1\left(\tilde{E}_{g(u)}^{B_{1g}}\right)$	$\psi_1^{s(t)} - i\psi_{14}^{s(t)} - \psi_4^{s(t)} + i\psi_{15}^{s(t)}$	$(001)(s)$	(010)	$(110), (\bar{1}10)$
$\Psi_2\left(\tilde{E}_{g(u)}^{B_{1g}}\right)$	$\psi_{27}^{s(t)} + i\psi_{37}^{s(t)} - \psi_{26}^{s(t)} - i\psi_{40}^{s(t)}$	$(001)(s)$		
$\Psi_1\left(\tilde{E}_{g(u)}^{B_{2g}}\right)$	$\psi_1^{s(t)} - i\psi_{14}^{s(t)} - \psi_4^{s(t)} + i\psi_{15}^{s(t)}$	$(001)(s)$	(100)	$(110), (\bar{1}10)$
$\Psi_2\left(\tilde{E}_{g(u)}^{B_{2g}}\right)$	$-\psi_{27}^{s(t)} - i\psi_{37}^{s(t)} + \psi_{26}^{s(t)} - i\psi_{40}^{s(t)}$	$(001)(s)$		

[1)] Nodes due to cancellation of basis functions of one row (for singlet functions only)
[2)] Nodes or dips due to interference of basis functions of two rows

Nodal structure of basis functions on planes may be analyzed by the same way as it was done for real functions. By acting with the reflection $h_{28} = Ih_4$ in plane (001) on the function of the first row, we obtain

$$\begin{aligned} Ih_4\Psi_1\left(\tilde{E}_{g(u)}\right) &= Ih_4\left(\psi_1^{s(t)} + i\psi_{14}^{s(t)} - \psi_4^{s(t)} - i\psi_{15}^{s(t)}\right) = I\left(\psi_4^{s(t)} + i\psi_{15}^{s(t)} - \psi_1^{s(t)} - i\psi_{14}^{s(t)}\right) \\ &= \pm\left(\psi_4^{s(t)} + i\psi_{15}^{s(t)} - \psi_1^{s(t)} - i\psi_{14}^{s(t)}\right) \end{aligned}$$

where plus or minus sign corresponds to singlet or triplet case, respectively. Thus we obtained that singlet functions $\tilde{E}_g$ are nodal but triplet functions $\tilde{E}_u$ are nodeless in the basal plane. The same results are obtained for the function of the second line.

It is seen Fig. 4.7 that the functions of first row in Table 4.24 don't meet at any vertical plane. Hence, it follows that there are no vertical nodal planes of the first type. When acting with reflections in planes (100) and (010) on the sum of function of $\Psi_1\left(\tilde{E}_{g(u)}\right)+\Psi_2\left(\tilde{E}_{g(u)}\right)$ we get

$$=h_{26}\left(\psi_1^{s(t)}+i\psi_{14}^{s(t)}-\psi_4^{s(t)}-i\psi_{15}^{s(t)}+\psi_{27}^{s(t)}-i\psi_{37}^{s(t)}-\psi_{26}^{s(t)}+i\psi_{40}^{s(t)}\right)=$$
$$\psi_{26}^{s(t)}+i\psi_{37}^{s(t)}-\psi_{27}^{s(t)}-i\psi_{40}^{s(t)}+\psi_4^{s(t)}-i\psi_{14}^{s(t)}-\psi_1^{s(t)}+i\psi_{15}^{s(t)}$$

and

$$h_{27}\left(\psi_1^{s(t)}+i\psi_{14}^{s(t)}-\psi_4^{s(t)}-i\psi_{15}^{s(t)}+\psi_{27}^{s(t)}-i\psi_{37}^{s(t)}-\psi_{26}^{s(t)}+i\psi_{40}^{s(t)}\right)=$$
$$\psi_{27}^{s(t)}+i\psi_{40}^{s(t)}-\psi_{26}^{s(t)}-i\psi_{37}^{s(t)}+\psi_1^{s(t)}-i\psi_{15}^{s(t)}-\psi_4^{s(t)}+i\psi_{14}^{s(t)})$$

The reflection h_{26} changed the sign, but the reflection h_{27} didn't change the sign, hence it follows that the function $\tilde{E}_{g(u)}$ has nodes in plane (100) and is nodeless in plane (010).

The action of reflection in diagonal plane $(\bar{1}10)$ on total function $\Psi_1\left(\tilde{E}_{g(u)}\right)+\Psi_2\left(\tilde{E}_{g(u)}\right)$ results in

$$h_{37}\left(\psi_1^{s(t)}+i\psi_{14}^{s(t)}-\psi_4^{s(t)}-i\psi_{15}^{s(t)}+\psi_{27}^{s(t)}-i\psi_{37}^{s(t)}-\psi_{26}^{s(t)}+i\psi_{40}^{s(t)}\right)=$$
$$\psi_{37}^{s(t)}+i\psi_{27}^{s(t)}-\psi_{40}^{s(t)}-i\psi_{26}^{s(t)}+\psi_{15}^{s(t)}-i\psi_1^{s(t)}-\psi_{15}^{s(t)}+i\psi_4^{s(t)}$$

It is seen from this relation that the reflections in diagonal plane result in phase factors i or $-i$ and the interference of the functions on the plane will result in dips, but not nodes.

So far the phase in each sector in k-space was constant and was defined by matrix elements of IR. It is seen from Table 4.24 that the basis functions $\psi_{14}^{s(t)}$, $\psi_4^{s(t)}$ and $\psi_{15}^{s(t)}$ of the first row of $\tilde{E}_{g(u)}$ obtained by rotations with angle $\bar{\theta}=\pi/2,\pi$ and $3\pi/2$ acquire phase factors $i,-1$ and $-i$, respectively, i.e. correspond to magnetic quantum number $\bar{m}=1$. Basis functions of the second row acquire phase factors $-i,-1$ and i and correspond to magnetic quantum number $\bar{m}=-1$. In Ginzburg-Landau theory, the SOP is identified with the wavefunction of a Cooper pair which includes all pairs in the condensate. In the space-group approach, this general pair function is obtained when the electron wavevector k_1 runs in the sector $0\le\theta\le\pi/4$ (Fig. 4.7) and other sectors are brought about by the elements of the left cosets. Suppose that there is a phase winding $\exp(im\theta)$ in the first sector. In order to obtain coinciding values of the angular momentum of a pair at discrete values of the angle $\bar{\theta}$, which is a property of an individual pair, and the angular momentum in the condensate, we assume that $m=\bar{m}$.

In axial symmetry case, the direction of k-vector depends on the azimuthal angle θ and k_z−projection and we consider θ- dependence of pair function and fix some nonzero k_z-value. This dependence may be included in general formula for pair function replacing k by $|k|$, θ and introducing phase winding in the first sector as follows:

$$\psi^{s(t)}_{|k|,\theta} = \left[\varphi_k(\xi_1)\varphi_{-k}(\xi_2) \pm \varphi_k(\xi_2)\varphi_{-k}(\xi_1)\right] \exp(im\theta) = \psi^{s(t)}_{|k|} \exp(im\theta), 0 \leq \theta \leq \pi/4 \quad (4.10.2)$$

Thereby the phase winding in the first sector is defined by topological phase winding quantum number m and the phase changing by discrete rotations is defined by quantum number $\bar{m}$, connected with IRs by formula (4.10.1).

According to Ginzburg-Landau theory, the wavefunction of a Cooper pair is identified with a SOP in condensate and a gauge phase winding may take place for all values of θ. Thus in order to obtain the same behavior of the pair function for all θ-values, namely, discrete and continuous rotations we should equate m in formula (4.10.2) to $\bar{m}$ in (4.10.1). By this means we obtain the functions with the same angular momentum in all odd sectors of Fig. 4.13a.

Consider the result of action with reflections σ_y on two functions $\psi^{s(t)}_{|k|} \exp(im\beta)$ and $\psi^{s(t)}_{|k|} \exp(im\alpha)$, defined by formula (4.10.2) with angles $\beta > \alpha$ and $m \geq 1$. The reflection transforms these functions into angles $-\chi$ and $-\eta$, with the unchanged phases. Thus when the angle θ moves from smaller value α to a larger value β, its image in 8th sector moves from larger value $-\alpha$ to a smaller value $-\beta$ with the same phase changing from $\exp(im\alpha)$ to $\exp(im\beta)$. Hence, it follows that the reflection changes the phase winding direction. It is clear that since multiplication by a group-theoretical phase factor $D^k(\sigma_y)$ is independent on angle θ, it doesn't change a phase. Thus we obtained that the reflection changes the phase winding direction, i.e. magnetic quantum number changes from m to $-m$.

The reversal of the sign of the quasi-angular momentum $\bar{m}$ upon reflection also follows from the form of the basis functions presented in Table 4.24. Indeed the function of the second row is obtained by reflection h_{27} and the $\bar{m}-$value in the second row equals to -1. Phase winding directions are indicated in Fig. 4.13 by arrows.

Basis functions of two rows of two-dimensional IRs are localized at different sectors in k-space (see Fig. 4.7) and meet at vertical planes where interference of pair functions occurs. In order to envisage such interference the wavefunctions of pairs in a finite number of points were represented by normalized sum of real and imaginary Gaussians. This numerical representation is possible for the cases with phase winding in the first sector (see formula (4.10.2)) and without phase winding. Within the sectors the phases at neighboring points coincide (if $m = 0$) or almost coincide (if $m \neq 0$) and constructive interference takes place. At the boundaries of the sectors, both constructive and destructive interferences of the wavefunctions are possible. When summing real and imaginary parts and squaring modulus of the complex function at every point we obtain the structure of SOP in the plane normal to k_z-axis. Nodal structures of IRs $\tilde{E}_{g(u)}$ and $\tilde{E}^{A_{2g}}_{g(u)}$ obtained by this numerical method presented in Fig. 4.13b, c. The nodal structures are defined by the interference on planes and are the same in the cases of phase winding and without phase winding. Both structures have

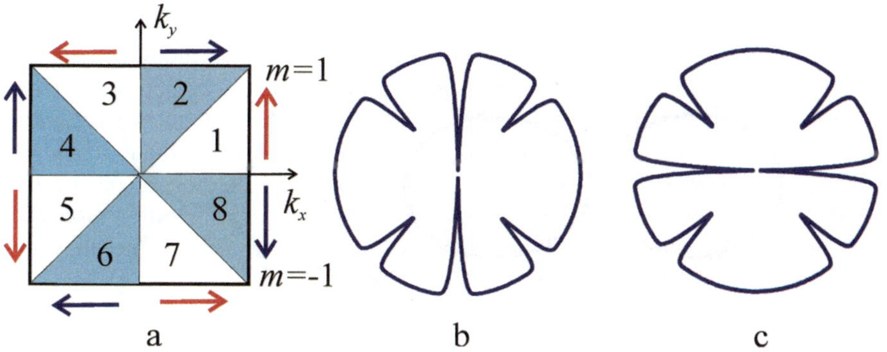

Fig. 4.13 Structure of chiral functions $\tilde{E}_{g(u)}$ in k-space in plane normal to k_z-direction. a white area—function of the first row, shadowed area—function of the second row, arrows indicate phase winding directions b-nodal structure of $\tilde{E}_{g(u)}$, c-nodal structure of $\tilde{E}^{A_{2g}}_{g(u)}$ (see Table 4.24). Note that $\tilde{E}_g$ is nodal and $\tilde{E}_u$ is nodeless in $k_z = 0$ plane

dips in diagonal planes, and also $\tilde{E}_{g(u)}$ has nodes in plane (100) and $\tilde{E}^{A_{2g}}_{g(u)}$ nodes in plane (010). Multiplication by A_{2g} changes the nodal structure in vertical coordinate planes leaving unchanged dips in diagonal planes. Nodal structures of $\tilde{E}^{B_{1g}}_{g(u)}$ and $\tilde{E}^{B_{2g}}_{g(u)}$ are the same as that of $\tilde{E}_{g(u)}$ and $\tilde{E}^{A_{2g}}_{g(u)}$, respectively, but phase winding directions are opposite to that shown in Fig. 4.13a. Since the values of m are positive in the half of sectors and are negative in the other half of the sectors these functions have zero total phase winding and are non-unitary.

Experimental data on SOP of Sr_2RuO_4 manifest transition between unitary and non-unitary SOP. Basis functions belonging to real IRs considered in a previous section are unitary. Also unitary states may be obtained as a sum of two non-unitary states with opposite phase winding direction in each sector. Figure 4.14 shows two types of the interference between the states with opposite angular momemtum projections. Figure 4.14a shows the sum of $\tilde{E}_{g(u)}$ and $\tilde{E}^{B_{1g}}_{g(u)}$ and Fig. 4.14b shows the sum of $\tilde{E}^{A_{2g}}_{g(u)}$ and $\tilde{E}^{B_{2g}}_{g(u)}$ with additional phase π. First structure is real and the second structure is imaginary in all sectors, thus both structures are unitary. These structures resemble modulus of atomic p_x- and p_y-orbitals.

Thus even and odd two-dimensional complex IRs of group D_{4h} have one vertical nodal plane (100) or (010) and dips in all diagonal planes (110) and (−110) The nodal and nodeless planes are interchanged when multiplying by one-dimensional IR A_{2g} so the vertical nodal planes of $\tilde{E}_g$ and $\bar{E}_u$ are topologically unstable in the sense of Kobayashi et al. [42, 43], i.e. nodes can be removed by small perturbation (in our case when multiplying by even one-dimensional IR). The basis functions of $\tilde{E}_g$ and $\tilde{E}_u$ symmetry in chiral representations have angular momentum projection $\bar{m} = \pm 1$. The essential difference appears in the basal plane, where $\tilde{E}_u$ is nodeless and nodes of $\tilde{E}_g$ are in all cases, and these nodes are topologically stable.

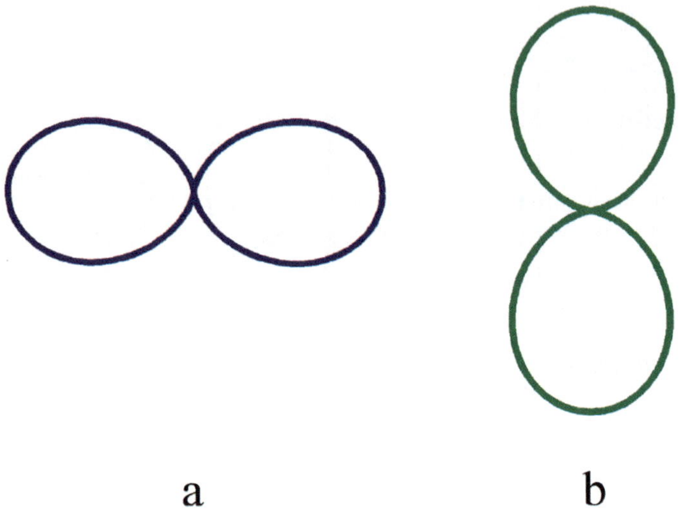

Fig. 4.14 The structure of SOP **a** $\tilde{E}_{g(u)} + \tilde{E}^{B_{1g}}_{g(u)}$ and **b** $\tilde{E}^{A_{2g}}_{g(u)} - \tilde{E}^{B_{2g}}_{g(u)}$

Going over to one-dimensional IRs we see in Table 4.9 that in characters of IRs $B_{1g(u)}$ and $B_{2g(u)}$ for rotations with angles $\pi/4$, π and $3/4\pi$ equal to -1, 1, -1 respectively, hence it follows that these IRs correspond to $\bar{m} = 2$, or to $\bar{m} = -2$. Figure 4.15 shows different possible structures of SOP for IRs B_{1g}, B_{2g}, B_{1u} and B_{2u}, calculated by the same numerical method with $\bar{m} = 2$ for functions of Tables 4.16 and 4.17. In the case of one-dimensional IR, the nodal structure is robust and nodes in Fig. 4.15 are the same as that in Tables 4.16 and 4.17. The structure of B_{1g} and B_{2u} in Fig. 4.15a has nodes in diagonal planes and the structure of B_{2g} and B_{1u} in Fig. 4.15c has nodes in vertical coordinate planes. Also B_{1u} and B_{2u} are nodal in the basal plane, but B_{1g} and B_{2g} are nodeless in the basal plane. Pair functions with $m = +2$ and with $m = -2$ are complex conjugate at all values of θ (see formula

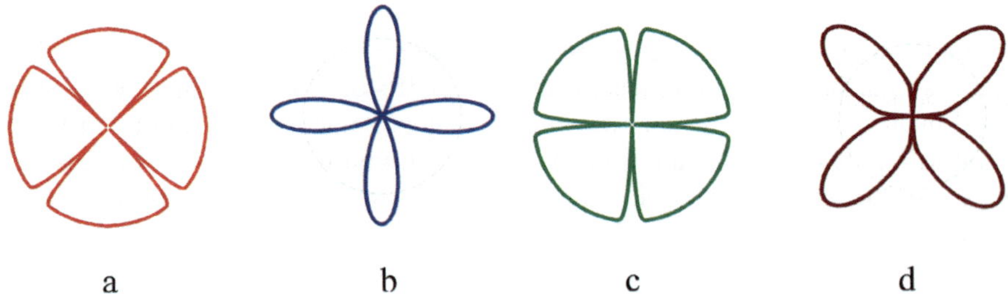

Fig. 4.15 Possible structures of SOP of $B-$symmetry in D_{4h} group. **a** B_{1g} and B_{2u}, $m = 2$, **b** B_{1g} and B_{2u}, interference of two waves $m = \pm 2$, **c** B_{2g} and B_{1u}, $m = 2$, **d** B_{2g} and B_{1u}, interference of two waves $m = \pm 2$

(4.10.2)). Figure 4.15b, d shows unitary linear combinations of B_{1g} ($m = \pm 2$) and B_{2g} ($m = \pm 2$), respectively. These structures resemble modulus of orbitals $d_{x^2-y^2}$ and d_{xy}, respectively, i.e. atomic $d-$orbitals in a crystal field.

4.11 Pair Functions in D_{2h} and $m'm'm$ Symmetry

Group-theoretical analysis, based on Mackey and Frobenius theorems, of possible spatial parts of electron pairs for the symmetry planes and lines in a BZ for the space group $Immm$ (point group D_{2h}) is presented in Table 4.25. Since multiplication table for the group D_{2h} is obvious, in present section, we use standard notations for point group elements. The data of this table can be easily obtained as follows. On symmetry planes $H = \mathrm{C}_s$ and $\hat{M}_\alpha = H + I = \mathrm{C}_{2h}$. Since pair's function is invariant under the action of reflection in plane, singlet pairs belong to A_g and triplet pairs belong to B_u of group C_{2h}. Making use of Frobenius reciprocity theorem one easily obtains pairs' symmetry on the whole group, obtained by induction of A_g and B_u for singlet and triplet pairs, respectively. In symmetry lines, wavevector group $H = \mathrm{C}_{2v}$ and characters $P^\pm$ don't depend on small IR and equal to 1 on C_{2v}. The extended group $\hat{M}_\alpha = H + IH$ is the whole group D_{2h} and characters of P^+ equal to 1 and characters of P^- equal to -1 for the elements of left coset IH. Consider, for example, IR B_{2u}. At a general point of a BZ all odd IRs are possible for spatial part of triplet pairs. It is seen from this table that B_{2u} has nodes on plane (010) and in directions k_x and k_z. Intersection of a plane with a FS results in a nodal line and intersection of line with a FS results in a point node.

Basis set for two-electron states in D_{2h} symmetry, shown in Fig. 4.16, consists of four functions defined by identity element, rotation C_{2z} and two reflections σ_x and σ_y. It is assumed that initial one-electron $k-$vector has some k_z-projection. Total

Table 4.25 Symmetry of spatial parts of singlet and triplet pairs on symmetry planes and lines, and possible symmetry on the whole group D_{2h} (two last columns). For the lines only characters of $P^\pm$ on H are presented

Planes	$M_\alpha = \mathrm{C}_{2h}$				IR (singlet)	IR (triplet)
(100)	E	σ_x	I	C_{2x}	A_g, B_{3g}	B_{1u}, B_{2u}
(010)	E	σ_y	I	C_{2y}	A_g, B_{2g}	B_{1u}, B_{3u}
(001)	E	σ_z	I	C_2	A_g, B_{1g}	B_{2u}, B_{3u}
$P^\pm$	1	1	± 1	± 1		
Lines	$H = \mathrm{C}_{2v}$				IR (singlet)	IR(triplet)
k_x	E	C_{2x}	σ_y	σ_z	A_g	B_{3u}
k_y	E	C_{2y}	σ_x	σ_z	A_g	B_{2u}
k_z	E	C_{2z}	σ_x	σ_y	A_g	B_{1u}
$P^\pm$	1	1	1	1[1)]		

[1)] Characters of P^+ and P^- on the left coset IH equal to 1 and -1, respectively

set of two-electron states on a FS is obtained if k_z runs in the interval $0 < k_z < b_3/2$ and the angle θ between k- and k_x-axis runs in the interval $0 < \theta < \pi/2$.

In D_{2h} symmetry, three triplet spin components $\hat{x}$, $\hat{y}$ and $\hat{z}$ are transformed independently by IRs B_{3g}, B_{2g} and B_{1g}, respectively, and the dimension of complete space of triplet two-electron states consists of 12 elements. Making use of projection operator technique one easily obtains triplet basis functions presented in Table 4.26.

Since these functions are expressed in terms of products of spatial parts and spin parts, their symmetry on the planes is B_u of C_{2h} if spin vector $\hat{r}_i$ is normal to the plane, and is A_u if $\hat{r}_i$ lies in the plane. Their nodal structure may be estimated acting by reflection operation on pair function. The function is nodeless on the plane, if it is invariant, and vice versa if it changes the sign, the function is nodal. The nodes are also indicated in Table 4.26.

The nodal structure of basis function on planes can be estimated by the action of reflection in these planes: if all signs change the function is nodal and if all signs are invariant, the function is nodeless. The spin components $\hat{x}$, $\hat{y}$ and $\hat{z}$ are transformed as vector components under rotations, but they are invariant under inversion. According to the basis choice, the action of the elements of C_{2v} induces permutation of basis functions without changing the sign. Since the reflection σ_z is not the element of C_{2v}, it is represented as IC_{2z}. The element C_{2z} induces permutation of basis function and inversion changes the sign of triplet function. As an example of application of these rules, we investigate nodes of the three functions of B_{1u}-type:

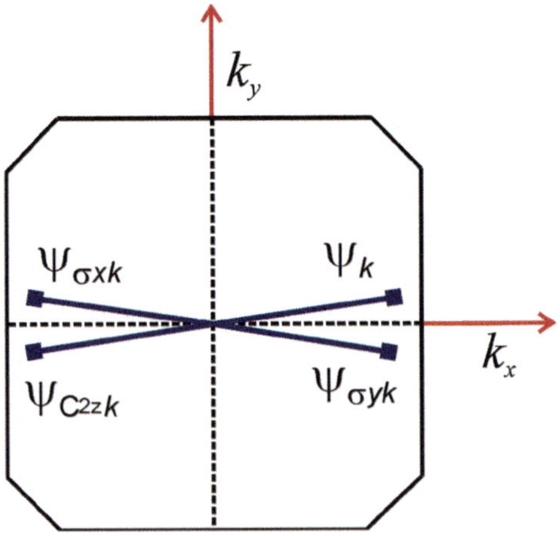

Fig. 4.16 Basis set of Cooper pairs in D_{2h} symmetry

Table 4.26 Triplet two-electron functions in D_{2h} symmetry and their nodal planes

Notation	Triplet functions	Nodes
$\Psi_{\hat{x}}\,(A_u)$	$\hat{x}\left(\psi_k^t - \psi_{C_{2z}k}^t - \psi_{\sigma_x k}^t + \psi_{\sigma_y k}^t\right)$	All planes
$\Psi_{\hat{y}}\,(A_u)$	$\hat{y}\left(\psi_k^t - \psi_{C_{2z}k}^t + \psi_{\sigma_x k}^t - \psi_{\sigma_y k}^t\right)$	All planes
$\Psi_{\hat{z}}\,(A_u)$	$\hat{z}\left(\psi_k^t + \psi_{C_{2z}k}^t + \psi_{\sigma_x k}^t + \psi_{\sigma_y k}^t\right)$	All planes
$\Psi_{\hat{x}}\,(B_{1u})$	$\hat{x}\left(\psi_k^t - \psi_{C_{2z}k}^t + \psi_{\sigma_x k}^t - \psi_{\sigma_y k}^t\right)$	(001)
$\Psi_{\hat{y}}\,(B_{1u})$	$\hat{y}\left(\psi_k^t - \psi_{C_{2z}k}^t - \psi_{\sigma_x k}^t + \psi_{\sigma_y k}^t\right)$	(001)
$\Psi_{\hat{z}}\,(B_{1u})$	$\hat{z}\left(\psi_k^t + \psi_{C_{2z}k}^t - \psi_{\sigma_x k}^t - \psi_{\sigma_y k}^t\right)$	(001)
$\Psi_{\hat{x}}\,(B_{2u})$	$\hat{x}\left(\psi_k^t + \psi_{C_{2z}k}^t + \psi_{\sigma_x k}^t + \psi_{\sigma_y k}^t\right)$	(010)
$\Psi_{\hat{y}}\,(B_{2u})$	$\hat{y}\left(\psi_k^t + \psi_{C_{2z}k}^t - \psi_{\sigma_x k}^t - \psi_{\sigma_y k}^t\right)$	(010)
$\Psi_{\hat{z}}\,(B_{2u})$	$\hat{z}(\psi_k - \psi_{C_{2z}k} - \psi_{\sigma_x k} + \psi_{\sigma_y k})$	(010)
$\Psi_{\hat{x}}\,(B_{3u})$	$\hat{x}(\psi_k + \psi_{C_{2z}k} - \psi_{\sigma_x k} - \psi_{\sigma_y k})$	(100)
$\Psi_{\hat{y}}\,(B_{3u})$	$\hat{y}(\psi_k + \psi_{C_{2z}k} + \psi_{\sigma_x k} + \psi_{\sigma_y k})$	(100)
$\Psi_{\hat{z}}\,(B_{3u})$	$\hat{z}(\psi_k - \psi_{C_{2z}k} + \psi_{\sigma_x k} - \psi_{\sigma_y k})$	(100)

$$\sigma_x \Psi_{\hat{x}}\,(B_{1u}) = \sigma_x \hat{x}\left(\psi_k^t - \psi_{C_{2z}k}^t + \psi_{\sigma_x k}^t - \psi_{\sigma_y k}^t\right) = \hat{x}\left(\psi_{\sigma_x k}^t - \psi_{\sigma_y k}^t + \psi_k^t - \psi_{C_{2z}k}^t\right) = \Psi_{\hat{x}}\,(B_{1u})$$

$$\sigma_y \Psi_{\hat{x}}\,(B_{1u}) = \sigma_y \hat{x}\left(\psi_k^t - \psi_{C_{2z}k}^t + \psi_{\sigma_x k}^t - \psi_{\sigma_y k}^t\right) = -\hat{x}\left(\psi_{\sigma_y k}^t - \psi_{\sigma_x k}^t + \psi_{C_{2z}k}^t - \psi_k^t\right) = \Psi_{\hat{x}}\,(B_{1u})$$

$$IC_{2z} \Psi_{\hat{x}}\,(B_{1u}) = IC_{2z} \hat{x}\left(\psi_k^t - \psi_{C_{2z}k}^t + \psi_{\sigma_x k}^t - \psi_{\sigma_y k}^t\right) = -\hat{x}\left(-\psi_{C_{2z}k}^t + \psi_k^t - \psi_{\sigma_y k}^t + \psi_{\sigma_x k}^t\right) = -\Psi_{\hat{x}}\,(B_{1u})$$

$$\sigma_x \Psi_{\hat{y}}\,(B_{1u}) = \sigma_x \hat{y}\left(\psi_k^t - \psi_{C_{2z}k}^t - \psi_{\sigma_x k}^t + \psi_{\sigma_y k}^t\right) = -\hat{y}\left(\psi_{\sigma_x k}^t - \psi_{\sigma_y k}^t - \psi_k^t + \psi_{C_{2z}k}^t\right) = \Psi_{\hat{y}}\,(B_{1u})$$

$$\sigma_y \Psi_{\hat{y}}\,(B_{1u}) = \sigma_y \hat{y}\left(\psi_k^t - \psi_{C_{2z}k}^t - \psi_{\sigma_x k}^t + \psi_{\sigma_y k}^t\right) = \hat{y}\left(\psi_{\sigma_y k}^t - \psi_{\sigma_x k}^t - \psi_{C_{2z}k}^t + \psi_k^t\right) = \Psi_{\hat{y}}\,(B_{1u})$$

$$IC_{2z} \Psi_{\hat{y}}\,(B_{1u}) = IC_{2z} \hat{y}\left(\psi_k^t - \psi_{C_{2z}k}^t - \psi_{\sigma_x k}^t + \psi_{\sigma_y k}^t\right) = -\hat{y}\left(-\psi_{C_{2z}k}^t + \psi_k^t + \psi_{\sigma_y k}^t - \psi_{\sigma_x k}^t\right) = -\Psi_{\hat{y}}\,(B_{1u})$$

$$\sigma_x \Psi_{\hat{z}}\,(B_{1u}) = \sigma_x \hat{z}\left(\psi_k^t + \psi_{C_{2z}k}^t - \psi_{\sigma_x k}^t - \psi_{\sigma_y k}^t\right) = -\hat{z}\left(\psi_{\sigma_x k}^t + \psi_{\sigma_y k}^t - \psi_k^t - \psi_{C_{2z}k}^t\right) = \Psi_{\hat{z}}\,(B_{1u})$$

$$\sigma_y \Psi_{\hat{z}}\,(B_{1u}) = \sigma_y \hat{z}\left(\psi_k^t + \psi_{C_{2z}k}^t - \psi_{\sigma_x k}^t - \psi_{\sigma_y k}^t\right) = -\hat{z}\left(\psi_{\sigma_y k}^t + \psi_{\sigma_x k}^t - \psi_{C_{2z}k}^t - \psi_k^t\right) = \Psi_{\hat{z}}\,(B_{1u})$$

$$IC_{2z} \Psi_{\hat{z}}\,(B_{1u}) = IC_{2z} \hat{z}\left(\psi_k^t + \psi_{C_{2z}k}^t - \psi_{\sigma_x k}^t - \psi_{\sigma_y k}^t\right) = \hat{z}\left(\psi_{C_{2z}k}^t + \psi_k^t - \psi_{\sigma_y k}^t - \psi_{\sigma_x k}^t\right) = -\Psi_{\hat{z}}\,(B_{1u})$$

Thus all functions are invariant under reflections σ_x and σ_y, and change the signs under reflection σ_z. Hence, it follows that all these total pairs' functions have nodes in plane (001). The same result was obtained for spatial parts of B_{1u} symmetry in Table 4.25.

The real ESP (equal spin pairing) functions $\hat{x}$ and $\hat{y}$ are linear combinations of $M_S = \pm 1$ states. However to represent triplet spin component with $M_S = \pm 1$ linear combinations $\hat{x} \pm i\hat{y}$ are required. However, these functions are not transformed by any IR of symmetry group D_{2h}. Hence, it follows that we need to find a symmetry group for transformation of a functions $\hat{x} + i\hat{y}$ and $\hat{x} - i\hat{y}$ in the crystal symmetry D_{2h}. This can be done by consideration of time-reversal symmetry and magnetic (Shubnikov) groups.

Total symmetry of a crystal with a point group D_{2h} includes also time reversal ϑ, and it is a grey Shubnikov group:

$$M = D_{2h} + \vartheta D_{2h} \tag{4.11.1}$$

Its subgroups are black-and-white groups $M = H + AH$, where H is halving subgroup and A is antiunitary element. In the cases of the black and white Shubnikov groups $m'm'm'$ and mmm' time reversal ϑ is connected with space inversion I. Note that according to the Anderson ansatz [78] wavefunctions of electrons in a pair are connected by I in ESP case or by ϑ in OSP (opposite spin pairing) case and the operation θI connects two spin projections with the same k. Hence, it follows that groups which include the product ϑI are not appropriate for Cooper pairing. It remains just a group $m'm'm$ in which C_{2h} is a unitary subgroup and antiunitary elements are ϑC_{2x}, ϑC_{2y}, $\vartheta \sigma_x$ and $\vartheta \sigma_y$. We choose antiunitary element $A = \vartheta \sigma_y$ as a representative of antiunitary left coset. Note that group $m'm'm$ only corresponds to ferromagnetic case in D_{2h} point symmetry (see Table 7.7 in [19]).

When applying Herring criterion in the case of integer spin, we obtain

$$\sum_B \chi\left(B^2\right) = 4 \tag{4.11.2}$$

Thus ICRs (irreducible corepresentations) of this Shubnikov group belong to the type (a) and are obtained as follows. For unitary subgroup C_{2h}, they coincide with its IRs A_g, A_u, B_g or B_u.

$$\Delta^k(h) = D^k(h), \ h \in \mathsf{C}_{2h} \tag{4.11.3}$$

They are extended on non-unitary elements with the sign $+$ or $-$.

$$\Delta^{k\pm}(B) = \pm D^k(BA^{-1})N \tag{4.11.4}$$
$$NN^* = D(A^2) = 1 \tag{4.11.5}$$

When taking $\vartheta\sigma_y$ as a non-unitary left coset representative we obtain that N is the identity matrix. We denote ICRs by the symbols of IRs of the unitary subgroup with the superscript $+$ or $-$. Thus, the group $m'm'm$ has four odd ICRs A_u^+, A_u^-, B_u^+ and B_u^-. Characters of these ICRs coincide with the characters of the IRs B_{1u}, A_u, B_{3u} and B_{2u} of group D_{2h}, respectively. Characters are presented in Table 4.27. Characters of even ICRs may be constructed by a similar way.

In magnetic field spin triplet function has three components S_0^1 and $S_{\pm 1}^1$, where subscripts stand for magnetic quantum numbers. We use basis functions $S_{\pm 1}^1 =$

Table 4.27 Odd ICRs of $m'm'm$ Shubnikov group. ICRs belong to type (a) and are labeled by IRs of unitary subgroup C_{2h} and a superscript + or −, which corresponds to formula (4.11.4). Labels of IRs of D_{2h} with the same characters are given in the parenthesis

ICR(IR)	E	σ_z	I	C_{2z}	$\vartheta\sigma_y$	ϑC_{2x}	ϑC_{2y}	$\vartheta\sigma_x$
A_u^+ (B_{1u})	1	−1	−1	1	1	−1	−1	1
A_u^- (A_u)	1	−1	−1	1	−1	1	1	−1
B_u^+ (B_{3u})	1	1	−1	−1	1	1	−1	−1
B_u^- (B_{2u})	1	1	−1	−1	−1	−1	1	1

$\hat{x} \pm i\hat{y}$ and $S_0^1 = \hat{z}$ for ESP and OSP spin functions, respectively. In basal plane spin functions $S_{\pm 1}^1 = \hat{x} \pm i\hat{y}$ belong to IR B_g of C_{2h}, spatial part belongs to B_u and total function belongs to A_u. The extension of A_u to $m'm'm$ results in ICRs A_u^+ and A_u^-.

Symmetry group of two vertical planes (100) and (010) is Shubnikov group $2'/m'$. Transformation properties of triplet spin functions on vertical symmetry planes are as follows:

$$\vartheta\sigma_x\left(\hat{x} + i\hat{y}\right) = \vartheta\left(\hat{x} - i\hat{y}\right) = \left(\hat{x} + i\hat{y}\right) \tag{4.11.6}$$

$$\vartheta\sigma_y\left(\hat{x} + i\hat{y}\right) = \vartheta\left(-\hat{x} + i\hat{y}\right) = -\hat{x} - i\hat{y} \tag{4.11.7}$$

It is also obvious that transformation properties of spin-down component $\hat{x} - i\hat{y}$ are the same. Thus the non-unitary elements of this group don't change the triplet spin projection $M_S = 1$. The resulting phase factor may be +1 or −1, corresponding to nodeless or nodal planes, respectively. It is seen from formulas (4.11.6) and (4.11.7) that the function $\left(\hat{x} + i\hat{y}\right)$ is nodeless in plane (100) and nodal in plane (010). However, multiplication by a phase factor i changes the nodal structure, indeed:

$$\vartheta\sigma_x\left(i\hat{x} - \hat{y}\right) = \vartheta\left(i\hat{x} + \hat{y}\right) = -\left(i\hat{x} - \hat{y}\right) \tag{4.11.8}$$

$$\vartheta\sigma_y\left(i\hat{x} - \hat{y}\right) = \vartheta\left(-i\hat{x} - \hat{y}\right) = \left(i\hat{x} - \hat{y}\right) \tag{4.11.9}$$

Thus starting from the pair function with spin $\hat{x} + i\hat{y}$ at k in the initial sector in a BZ (Fig. 4.16) and acting by the elements $\vartheta\sigma_x$, $\vartheta\sigma_y$ and C_{2z} we obtain the total basis set. All terms of this basis set have the same projection $M_S = 1$. Shubnikov group $m'm'm$ and spin directions in sectors are shown in Fig. 4.17.

Basis functions of ICRs of magnetic group $m'm'm$ may be obtained by projection operators, which include time reversal and spin orbitals:

$$\Psi_i^{\Gamma^q} = \frac{|\Gamma^q|}{|\bar{G}'|} \sum_{h_j \in \bar{G}'} \Gamma_{ii}^q\left(h_j\right) h'_j S_\alpha^1 \psi_k^t \,, \tag{4.11.10}$$

where the prime of h' means that for non-unitary elements it is ϑh_i and S_α^1 stands for one of the triplet spin component $\hat{z}$ or $\hat{x} \pm i\hat{y}$.

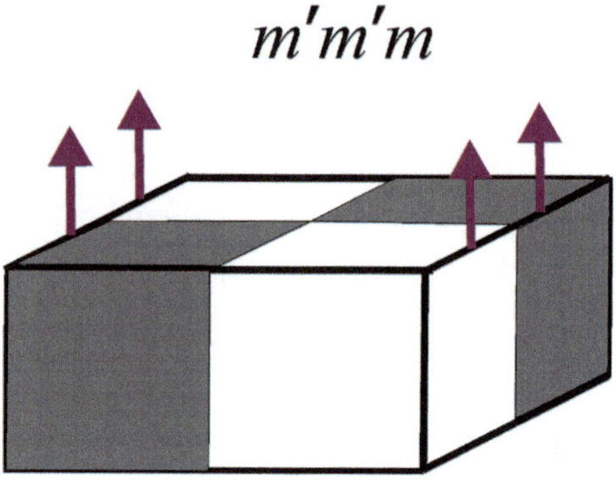

Fig. 4.17 Transformation of triplet spin component in black-and-white Shubnikov group $m'm'm$

Table 4.28 Triplet pair functions with $\bar{m} = \pm 1$ in $m'm'm$ symmetry

	Pair function	Nodal planes
$\Psi_{\hat{x}+i\hat{y}}(A_u^+)$	$(\hat{x}+i\hat{y})\left(\psi_k^t - \psi_{C_{2z}k}^t + \psi_{\sigma_x k}^t - \psi_{\sigma_y k}^t\right)$	(001)
$\Psi_{\hat{x}-i\hat{y}}(A_u^+)$	$(\hat{x}-i\hat{y})\left(\psi_k^t - \psi_{C_{2z}k}^t - \psi_{\sigma_x k}^t + \psi_{\sigma_y k}^t\right)$	All planes
$\Psi_{\hat{x}+i\hat{y}}(A_u^-)$	$(\hat{x}+i\hat{y})\left(\psi_k^t - \psi_{C_{2z}k}^t - \psi_{\sigma_x k}^t + \psi_{\sigma_y k}^t\right)$	All planes
$\Psi_{\hat{x}-i\hat{y}}(A_u^-)$	$(\hat{x}-i\hat{y})\left(\psi_k^t - \psi_{C_{2z}k}^t + \psi_{\sigma_x k}^t - \psi_{\sigma_y k}^t\right)$	(001)
$\Psi_{\hat{x}+i\hat{y}}(B_u^+)$	$(\hat{x}+i\hat{y})\left(\psi_k^t + \psi_{C_{2z}k}^t + \psi_{\sigma_x k}^t + \psi_{\sigma_y k}^t\right)$	(010)
$\Psi_{\hat{x}-i\hat{y}}(B_u^+)$	$(\hat{x}-i\hat{y})\left(\psi_k^t + \psi_{C_{2z}k}^t - \psi_{\sigma_x k}^t - \psi_{\sigma_y k}^t\right)$	(100)
$\Psi_{\hat{x}+i\hat{y}}(B_u^-)$	$(\hat{x}+i\hat{y})\left(\psi_k^t + \psi_{C_{2z}k}^t - \psi_{\sigma_x k}^t - \psi_{\sigma_y k}^t\right)$	(100)
$\Psi_{\hat{x}-i\hat{y}}(B_u^-)$	$(\hat{x}-i\hat{y})\left(\psi_k^t + \psi_{C_{2z}k}^t + \psi_{\sigma_x k}^t + \psi_{\sigma_y k}^t\right)$	(010)

In this section, we don't consider the phase winding of spatial parts. In this case, spatial part defined by (4.7.1) is real and invariant under complex conjugation and time reversal. Total triplet pairs' functions belonging to ICRs of $m'm'm$ magnetic group are presented in Table 4.28. Nodal structure of these functions depends on the behavior of spin and spatial part under the action of the elements $\vartheta\sigma_x$, $\vartheta\sigma_y$and σ_z. The action of elements σ_x and σ_x results in permutation of basis spatial functions without changing signs. Also σ_x changes the sign of $\hat{y}$ and σ_y changes the sign of $\hat{x}$, and ϑ results in complex conjugation. Since σ_z doesn't belong to C_{2v} it is represented as IC_{2z}. In this case C_{2z} permutes spatial basis functions without changing the signs and changes the signs of spin components, and space inversion changes the sign of triplet spatial parts.

The analysis of nodal structure of triplet pairs' functions of Table 4.28 is presented below:

$$\vartheta\sigma_x \Psi_{\hat{x}+i\hat{y}}(A_u^+) = \vartheta\sigma_x\left(\hat{x}+i\hat{y}\right)\left(\psi_k^t - \psi_{C_{2z}k}^t + \psi_{\sigma x k}^t - \psi_{\sigma y k}^t\right) =$$
$$= \vartheta\left(\hat{x}-i\hat{y}\right)\left(\psi_{\sigma x k}^t - \psi_{\sigma y k}^t + \psi_k^t - \psi_{C_{2z}k}^t\right) = \left(\hat{x}+i\hat{y}\right)\left(\psi_{\sigma x k}^t - \psi_{\sigma y k}^t + \psi_k^t - \psi_{C_{2z}k}^t\right) = \Psi_{\hat{x}+i\hat{y}}(A_u^+)$$
$$\vartheta\sigma_y \Psi_{\hat{x}+i\hat{y}}(A_u^+) = \vartheta\sigma_y\left(\hat{x}+i\hat{y}\right)\left(\psi_k^t - \psi_{C_{2z}k}^t + \psi_{\sigma x k}^t - \psi_{\sigma y k}^t\right) =$$
$$= \vartheta\left(-\hat{x}+i\hat{y}\right)\left(\psi_{\sigma y k}^t - \psi_{\sigma x k}^t + \psi_{C_{2z}k}^t - \psi_k^t\right) = -\left(\hat{x}+i\hat{y}\right)\left(\psi_{\sigma y k}^t - \psi_{\sigma x k}^t + \psi_{C_{2z}k}^t - \psi_k^t\right) = \Psi_{\hat{x}+i\hat{y}}(A_u^+).$$
$$IC_{2z}\Psi_{\hat{x}+i\hat{y}}(A_u^+) = IC_{2z}\left(\hat{x}+i\hat{y}\right)\left(\psi_k^t - \psi_{C_{2z}k}^t + \psi_{\sigma x k}^t - \psi_{\sigma y k}^t\right) =$$
$$\left(-\hat{x}-i\hat{y}\right)\left(-\psi_{C_{2z}k}^t + \psi_k^t - \psi_{\sigma y k}^t + \psi_{\sigma x k}^t\right) = -\Psi_{\hat{x}+i\hat{y}}(A_u^+)$$
$$\vartheta\sigma_x \Psi_{\hat{x}-i\hat{y}}(A_u^+) = \vartheta\sigma_x\left(\hat{x}-i\hat{y}\right)\left(\psi_k^t - \psi_{C_{2z}k}^t - \psi_{\sigma x k}^t + \psi_{\sigma y k}^t\right) =$$
$$\vartheta\left(\hat{x}+i\hat{y}\right)\left(\psi_{\sigma x k}^t - \psi_{\sigma y k}^t - \psi_k^t + \psi_{C_{2z}k}^t\right) = \left(\hat{x}-i\hat{y}\right)\left(\psi_{\sigma x k}^t - \psi_{\sigma y k}^t - \psi_k^t + \psi_{C_{2z}k}^t\right) = -\Psi_{\hat{x}-i\hat{y}}(A_u^+)$$
$$\vartheta\sigma_y \Psi_{\hat{x}-i\hat{y}}(A_u^+) = \vartheta\sigma_y\left(\hat{x}-i\hat{y}\right)\left(\psi_k^t - \psi_{C_{2z}k}^t - \psi_{\sigma x k}^t + \psi_{\sigma y k}^t\right) =$$
$$\vartheta\left(-\hat{x}-i\hat{y}\right)\left(\psi_{\sigma y k}^t - \psi_{\sigma x k}^t - \psi_{C_{2z}k}^t + \psi_k^t\right) = -\left(\hat{x}-i\hat{y}\right)\left(\psi_{\sigma y k}^t - \psi_{\sigma x k}^t - \psi_{C_{2z}k}^t + \psi_k^t\right) = -\Psi_{\hat{x}-i\hat{y}}(A_u^+)$$
$$IC_{2z}\Psi_{\hat{x}-i\hat{y}}(A_u^+) = IC_{2z}\left(\hat{x}-i\hat{y}\right)\left(\psi_k^t - \psi_{C_{2z}k}^t - \psi_{\sigma x k}^t + \psi_{\sigma y k}^t\right) =$$
$$\left(-\hat{x}+i\hat{y}\right)\left(-\psi_{C_{2z}k}^t + \psi_k^t + \psi_{\sigma y k}^t - \psi_{\sigma x k}^t\right) = -\Psi_{\hat{x}-i\hat{y}}(A_u^+)$$
$$\vartheta\sigma_x \Psi_{\hat{x}+i\hat{y}}(A_u^-) = \vartheta\sigma_x\left(\hat{x}+i\hat{y}\right)\left(\psi_k^t - \psi_{C_{2z}k}^t - \psi_{\sigma x k}^t + \psi_{\sigma y k}^t\right) =$$
$$\vartheta\left(\hat{x}-i\hat{y}\right)\left(\psi_{\sigma x k}^t - \psi_{\sigma y k}^t - \psi_k^t + \psi_{C_{2z}k}^t\right) = \left(\hat{x}+i\hat{y}\right)\left(\psi_{\sigma x k}^t - \psi_{\sigma y k}^t - \psi_k^t + \psi_{C_{2z}k}^t\right) = -\Psi_{\hat{x}+i\hat{y}}(A_u^-)$$
$$\vartheta\sigma_y \Psi_{\hat{x}+i\hat{y}}(A_u^-) = \vartheta\sigma_y\left(\hat{x}+i\hat{y}\right)\left(\psi_k^t - \psi_{C_{2z}k}^t - \psi_{\sigma x k}^t + \psi_{\sigma y k}^t\right) =$$
$$\vartheta\left(-\hat{x}+i\hat{y}\right)\left(\psi_{\sigma y k}^t - \psi_{\sigma x k}^t - \psi_{C_{2z}k}^t + \psi_k^t\right) = -\left(\hat{x}+i\hat{y}\right)\left(\psi_{\sigma y k}^t - \psi_{\sigma x k}^t - \psi_{C_{2z}k}^t + \psi_k^t\right) = -\Psi_{\hat{x}+i\hat{y}}(A_u^-)$$
$$IC_{2z}\Psi_{\hat{x}+i\hat{y}}(A_u^-) = IC_{2z}\left(\hat{x}+i\hat{y}\right)\left(\psi_k^t - \psi_{C_{2z}k}^t - \psi_{\sigma x k}^t + \psi_{\sigma y k}^t\right) =$$
$$\left(-\hat{x}-i\hat{y}\right)\left(-\psi_{C_{2z}k}^t + \psi_k^t + \psi_{\sigma y k}^t - \psi_{\sigma x k}^t\right) = -- \Psi_{\hat{x}+i\hat{y}}(A_u^-)$$
$$\vartheta\sigma_x \Psi_{\hat{x}-i\hat{y}}(A_u^-) = \vartheta\sigma_x\left(\hat{x}-i\hat{y}\right)\left(\psi_k^t - \psi_{C_{2z}k}^t + \psi_{\sigma x k}^t - \psi_{\sigma y k}^t\right) =$$
$$\vartheta\left(\hat{x}+i\hat{y}\right)\left(\psi_{\sigma x k}^t - \psi_{\sigma y k}^t + \psi_k^t - \psi_{C_{2z}k}^t\right) = \left(\hat{x}-i\hat{y}\right)\left(\psi_{\sigma x k}^t - \psi_{\sigma y k}^t + \psi_k^t - \psi_{C_{2z}k}^t\right) = +\Psi_{\hat{x}-i\hat{y}}(A_u^-)$$
$$\vartheta\sigma_y \Psi_{\hat{x}-i\hat{y}}(A_u^-) = \vartheta\sigma_y\left(\hat{x}-i\hat{y}\right)\left(\psi_k^t - \psi_{C_{2z}k}^t + \psi_{\sigma x k}^t - \psi_{\sigma y k}^t\right) =$$
$$\vartheta\left(-\hat{x}-i\hat{y}\right)\left(\psi_{\sigma y k}^t - \psi_{\sigma x k}^t + \psi_{C_{2z}k}^t - \psi_k^t\right) = -\left(\hat{x}-i\hat{y}\right)\left(\psi_{\sigma y k}^t - \psi_{\sigma x k}^t + \psi_{C_{2z}k}^t - \psi_k^t\right) = +\Psi_{\hat{x}-i\hat{y}}(A_u^-)$$
$$IC_{2z}\Psi_{\hat{x}-i\hat{y}}(A_u^-) = IC_{2z}\left(\hat{x}-i\hat{y}\right)\left(\psi_k^t - \psi_{C_{2z}k}^t + \psi_{\sigma x k}^t - \psi_{\sigma y k}^t\right) =$$
$$\left(-\hat{x}+i\hat{y}\right)\left(-\psi_{C_{2z}k}^t + \psi_k^t - \psi_{\sigma y k}^t + \psi_{\sigma x k}^t\right) = -\Psi_{\hat{x}-i\hat{y}}(A_u^-)$$
$$\vartheta\sigma_x \Psi_{\hat{x}+i\hat{y}}(B_u^+) = \vartheta\sigma_x\left(\hat{x}+i\hat{y}\right)\left(\psi_k^t + \psi_{C_{2z}k}^t + \psi_{\sigma x k}^t + \psi_{\sigma y k}^t\right) =$$
$$\vartheta\left(\hat{x}-i\hat{y}\right)\left(\psi_{\sigma x k}^t + \psi_{\sigma y k}^t + \psi_k^t + \psi_{C_{2z}k}^t\right) = \left(\hat{x}+i\hat{y}\right)\left(\psi_{\sigma x k}^t + \psi_{\sigma y k}^t + \psi_k^t + \psi_{C_{2z}k}^t\right) = +\Psi_{\hat{x}+i\hat{y}}(B_u^+)$$
$$\vartheta\sigma_y \Psi_{\hat{x}+i\hat{y}}(B_u^+) = \vartheta\sigma_y\left(\hat{x}+i\hat{y}\right)\left(\psi_k^t + \psi_{C_{2z}k}^t + \psi_{\sigma x k}^t + \psi_{\sigma y k}^t\right) =$$
$$\vartheta\sigma_y\left(-\hat{x}+i\hat{y}\right)\left(\psi_{\sigma y k}^t + \psi_{\sigma x k}^t + \psi_{C_{2z}k}^t + \psi_k^t\right) = -\left(\hat{x}+i\hat{y}\right)\left(\psi_{\sigma y k}^t + \psi_{\sigma x k}^t + \psi_{C_{2z}k}^t + \psi_k^t\right) = -\Psi_{\hat{x}+i\hat{y}}(B_u^+)$$
$$IC_{2z}\Psi_{\hat{x}+i\hat{y}}(B_u^+) = IC_{2z}\left(\hat{x}+i\hat{y}\right)\left(\psi_k^t + \psi_{C_{2z}k}^t + \psi_{\sigma x k}^t + \psi_{\sigma y k}^t\right) =$$
$$\left(-\hat{x}-i\hat{y}\right)\left(\psi_{C_{2z}k}^t + \psi_k^t + \psi_{\sigma y k}^t + \psi_{\sigma x k}^t\right) = +\Psi_{\hat{x}+i\hat{y}}(B_u^+)$$

$$\vartheta\sigma_x\Psi_{\hat{x}-i\hat{y}}(B_u^+) = \vartheta\sigma_x\,(\hat{x}-i\hat{y})\left(\psi_k^t+\psi_{C_{2z}k}^t-\psi_{\sigma_x k}^t-\psi_{\sigma_y k}^t\right) =$$
$$\vartheta\,(\hat{x}+i\hat{y})\left(\psi_{\sigma_x k}^t+\psi_{\sigma_y k}^t-\psi_k^t-\psi_{C_{2z}k}^t\right) = (\hat{x}-i\hat{y})\left(\psi_{\sigma_x k}^t+\psi_{\sigma_y k}^t-\psi_k^t-\psi_{C_{2z}k}^t\right) = -\Psi_{\hat{x}-i\hat{y}}(B_u^+)$$
$$\vartheta\sigma_y\Psi_{\hat{x}-i\hat{y}}(B_u^+) = \vartheta\sigma_y\,(\hat{x}-i\hat{y})\left(\psi_k^t+\psi_{C_{2z}k}^t-\psi_{\sigma_x k}^t-\psi_{\sigma_y k}^t\right) =$$
$$\vartheta\sigma_y\,(-\hat{x}-i\hat{y})\left(\psi_{\sigma_y k}^t+\psi_{\sigma_x k}^t-\psi_{C_{2z}k}^t-\psi_k^t\right) = -(\hat{x}-i\hat{y})\left(\psi_{\sigma_y k}^t+\psi_{\sigma_x k}^t-\psi_{C_{2z}k}^t-\psi_k^t\right) = \Psi_{\hat{x}-i\hat{y}}(B_u^+)$$
$$IC_{2z}\Psi_{\hat{x}-i\hat{y}}(B_u^+) = IC_{2z}\,(\hat{x}-i\hat{y})\left(\psi_k^t+\psi_{C_{2z}k}^t-\psi_{\sigma_x k}^t-\psi_{\sigma_y k}^t\right) =$$
$$(-\hat{x}+i\hat{y})\left(-\psi_{C_{2z}k}^t-\psi_k^t+\psi_{\sigma_y k}^t+\psi_{\sigma_x k}^t\right) = +\Psi_{\hat{x}-i\hat{y}}(B_u^+)$$
$$\vartheta\sigma_x\Psi_{\hat{x}+i\hat{y}}(B_u^-) = \vartheta\sigma_x\,(\hat{x}+i\hat{y})\left(\psi_k^t+\psi_{C_{2z}k}^t-\psi_{\sigma_x k}^t-\psi_{\sigma_y k}^t\right) =$$
$$\vartheta\,(\hat{x}-i\hat{y})\left(\psi_{\sigma_x k}^t+\psi_{\sigma_y k}^t-\psi_k^t-\psi_{C_{2z}k}^t\right) = (\hat{x}+i\hat{y})\left(\psi_{\sigma_x k}^t+\psi_{\sigma_y k}^t-\psi_k^t-\psi_{C_{2z}k}^t\right) = -\Psi_{\hat{x}+i\hat{y}}(B_u^-)$$
$$\vartheta\sigma_y\Psi_{\hat{x}+i\hat{y}}(B_u^-) = \vartheta\sigma_y\,(\hat{x}+i\hat{y})\left(\psi_k^t+\psi_{C_{2z}k}^t-\psi_{\sigma_x k}^t-\psi_{\sigma_y k}^t\right) =$$
$$= \vartheta\,(-\hat{x}+i\hat{y})\left(\psi_{\sigma_y k}^t+\psi_{\sigma_x k}^t-\psi_{C_{2z}k}^t-\psi_k^t\right) = -(\hat{x}+i\hat{y})\left(\psi_{\sigma_y k}^t+\psi_{\sigma_x k}^t-\psi_{C_{2z}k}^t-\psi_k^t\right) = \Psi_{\hat{x}+i\hat{y}}(B_u^-)$$
$$IC_{2z}\Psi_{\hat{x}+i\hat{y}}(B_u^-) = IC_{2z}\,(\hat{x}+i\hat{y})\left(\psi_{\sigma_y k}^t+\psi_{\sigma_x k}^t-\psi_{C_{2z}k}^t-\psi_k^t\right) =$$
$$= IC_{2z}\,(-\hat{x}-i\hat{y})\left(-\psi_{\sigma_x k}^t-\psi_{\sigma_y k}^t+\psi_k^t+\psi_{C_{2z}k}^t\right) = \Psi_{\hat{x}+i\hat{y}}(B_u^-)$$
$$\vartheta\sigma_x\Psi_{\hat{x}-i\hat{y}}(B_u^-) = \vartheta\sigma_x\,(\hat{x}-i\hat{y})\left(\psi_k^t+\psi_{C_{2z}k}^t+\psi_{\sigma_x k}^t+\psi_{\sigma_y k}^t\right) =$$
$$\vartheta\,(\hat{x}+i\hat{y})\left(\psi_{\sigma_x k}^t+\psi_{\sigma_y k}^t+\psi_k^t+\psi_{C_{2z}k}^t\right) = (\hat{x}-i\hat{y})\left(\psi_{\sigma_x k}^t+\psi_{\sigma_y k}^t+\psi_k^t+\psi_{C_{2z}k}^t\right) = +\Psi_{\hat{x}-i\hat{y}}(B_u^-)$$
$$\vartheta\sigma_y\Psi_{\hat{x}-i\hat{y}}(B_u^-) = \vartheta\sigma_y\,(\hat{x}-i\hat{y})\left(\psi_k^t+\psi_{C_{2z}k}^t+\psi_{\sigma_x k}^t+\psi_{\sigma_y k}^t\right) =$$
$$\vartheta\,(-\hat{x}-i\hat{y})\left(\psi_{\sigma_y k}^t+\psi_{\sigma_x k}^t+\psi_{C_{2z}k}^t+\psi_k^t\right) = (-\hat{x}+i\hat{y})\left(\psi_{\sigma_y k}^t+\psi_{\sigma_x k}^t+\psi_{C_{2z}k}^t+\psi_k^t\right) = -\Psi_{\hat{x}-i\hat{y}}(B_u^-)$$
$$IC_{2z}\Psi_{\hat{x}-i\hat{y}}(B_u^-) = IC_{2z}\,(\hat{x}-i\hat{y})\left(\psi_k^t+\psi_{C_{2z}k}^t+\psi_{\sigma_x k}^t+\psi_{\sigma_y k}^t\right) =$$
$$IC_{2z}\,(-\hat{x}+i\hat{y})\left(-\psi_{C_{2z}k}^t-\psi_k^t-\psi_{\sigma_y k}^t-\psi_{\sigma_x k}^t\right) = +\Psi_{\hat{x}-i\hat{y}}(B_u^-)$$

The cases when the reflection σ_z or reflections with time reversal $\vartheta\sigma_x$or $\vartheta\sigma_y$ changes the sign of total pair's function correspond to nodal planes, indicated in Table 4.28. It also follows from the above analysis that these operations don't change the pair's triplet spin projection.

4.12 Phase Winding in Magnetic Symmetry

Superconductivity is a manifestation of broken symmetry in nature. Spontaneously broken gauge symmetry $U(1)$ means that below T_c the wavefunction of the system spontaneously develops a definite phase, which can be treated as an additional degree of freedom. The quantum mechanical phase of one wavefunction itself is not a physical observable. However, the phase difference between two or more wavefunctions results in interference effects, which may be observable. The connection of phase winding with point group operations was considered in Sect. 4.10, and it was shown that in point group D_{4h} phase winding directions in adjacent sectors in $k-$ space are opposite and total phase winding vanished. In the present section, the technique is extended on Shubnikov point groups and it is shown that in order to obtain total phase winding $2\pi m$ the reflections in vertical mirror planes should be connected with time reversal ϑ.

According to the space-group theory, the wavevector runs over the basis domain of a BZ, whose volume is $|\bar{G}|$ times less than the volume of the whole BZ [19]. The wavevectors and wavefunctions in the other domains are obtained by the action with point group $\bar{G}$ elements on wavevector k and wavefunction ψ_k in the basis domain. Since the space inversion is included in the definition of a Cooper pair, the dimension of basis is reduced to $|\bar{G}|/2$. Each basis domain for the pairs includes two basis domains for representations connected by inversion. To obtain possible winding numbers for representations of D_{nh} group, it is sufficient to consider C_{nh} group. With this approach, a relation between the representations of the cyclic group and the possible angular momentums in the basic domain of a BZ can be obtained from the one-dimensional IRs D^k of cyclic group by equating the angular momentum m of the phase winding in the first sector to the quasi-angular momentum $\bar{m}$ for rotations by discrete angles $\bar{\theta}$ (4.10.1).

In the case of D_{2h} group, there is only one rotation element C_{2z}. It is seen from Table 4.26 that for D_{2h} group $A_u(C_{2z}) = +1$ $B_{1u}(C_{2z}) = +1$, $B_{2u}(C_{2z}) = -1$, and $B_{3u}(C_{2z}) = -1$. Hence, it follows that the values $\bar{m} = \pm 2$ may be associated with A_u and B_{1u} and the values $\bar{m} = \pm 1$ may be associated with B_{2u} and B_{3u}.

The phase windings with $m = 2$ for A_u and B_{1u} and with $m = 1$ for B_{2u} and B_{3u} of D_{2h} group are represented in Fig. 4.18a–d by arrows in a plane normal to (001) direction. If the initial k_1 vector runs over the first sector we obtain all pairs in condensate when acting by the elements of group C_{2v} on k_1. Consider the case, when the initial vector k_1 is close to k_x direction, and its phase is infinitesimal, thus the value of the pair function ψ_1^t is positive. The other $k-$vectors are obtained by the action of group elements as follows:

$$k_2 = \sigma_x k_1, \quad k_3 = C_{2z} k_1, \quad k_4 = \sigma_y k_1 \tag{4.12.1}$$

Basis functions obtained by Formula 4.8.5 are presented in Table 4.29. The phase winding in sectors 1 and 3 is defined by formula 4.10.2 in interval $0 < \vartheta < \pi/2$ and in sectors 2 and 4 the phase winding is opposite to that in sectors 1 and 3.

Consider the phase structure of B_{2u} shown in (Fig. 4.18c). The wavefunction of a pair can have a phase winding in the sector 1 corresponding to $m = 1$. Thus in the sector 1 the phase changes from 0 to $\pi/2$. When acting on function ψ_1 by the group element C_{2z} and multiplying by $B_{2u}(C_{2z}) = -1$ we obtain phase π at the beginning of the sector 3 (at k_3) and phase $3\pi/2$ at the end of this sector (near $-k_y$). Since $B_{2u}(\sigma_x) = 1$ the phase at k_2 equals to 0 and since $B_{2u}(\sigma_y) = -1$, phase at k_4 equals to π. Since reflection reverses phase winding direction, the phase winding in sectors 2 and 4 corresponds to $m = -1$. Thus, we obtain the phase structure shown in Fig. 4.18c. It is seen in Fig. 4.18c that the phase difference between functions of neighboring sectors near k_x-axis equals to π, and there are nodes at k_x-axis (at plane (010) in three dimensions), which are denoted by bold red line. Since the signs

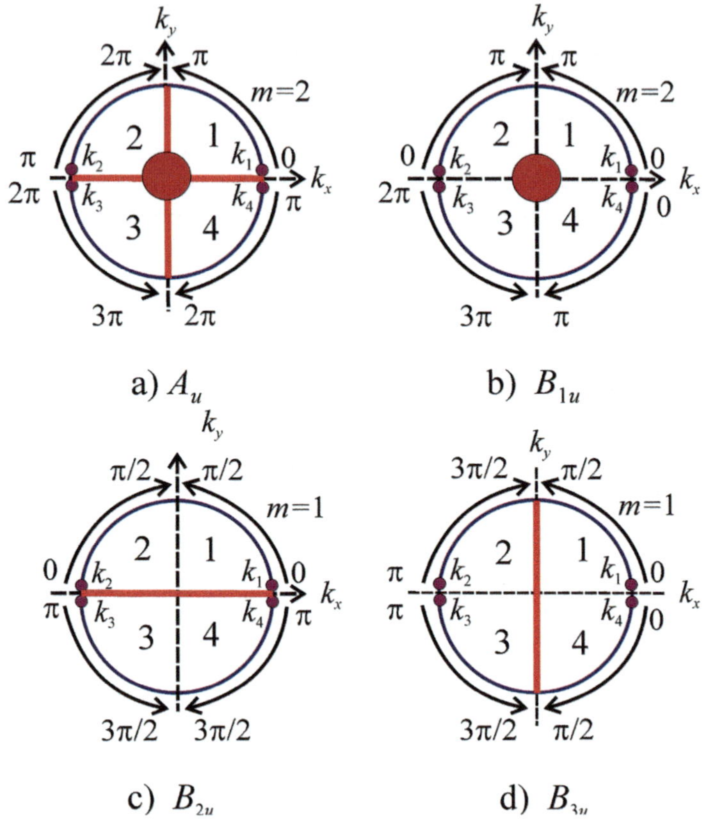

Fig. 4.18 Nodal structure and phase winding of SOP in D_{2h} symmetry for odd IRs **a** A_u, **b** B_{1u}, **c** B_{2u} and **d** B_{3u}. Bold red lines denote vertical nodal planes and red circles denote nodal basal plane

Table 4.29 Triplet basis functions with phase winding in D_{2h} symmetry

IR	Function	Nodal planes
$\Psi(A_{1u})$	$\psi^t_{k_1}e^{i2\theta} - \psi^t_{\sigma_x k_1}e^{-i2\theta} + \psi^t_{C_{2z}k_1}e^{i2\theta} - \psi^t_{\sigma_y k_1}e^{-i2\theta}$	All planes
$\Psi(B_{1u})$	$\psi^t_{k_1}e^{i2\theta} + \psi^t_{\sigma_x k_1}e^{-i2\theta} + \psi^t_{C_{2z}k_1}e^{i2\theta} + \psi^t_{\sigma_y k_1}e^{-i2\theta}$	(001)
$\Psi(B_{2u})$	$\psi^t_{k_1}e^{i\theta} + \psi^t_{\sigma_x k_1}e^{-i\theta} - \psi^t_{C_{2z}k_1}e^{i\theta} - \psi^t_{\sigma_y k_1}e^{-i\theta}$	(010)
$\Psi(B_{3u})$	$\psi^t_{k_1}e^{i\theta} - \psi^t_{\sigma_x k_1}e^{-i\theta} - \psi^t_{C_{2z}k_1}e^{i\theta} + \psi^t_{\sigma_y k_1}e^{-i\theta}$	(100)

of B_{3u} and B_{2u} are opposite for the reflections σ_x and σ_y, the nodes of B_{3u} in Fig. 4.18d are in k_y-direction (plane (100) in three dimensions).

Nodal structure of A_u and B_{1u} shown in Fig. 4.18a, b can be obtained by a similar way with $m = 2$. IR A_u has nodes in both vertical planes, but B_{1u} is nodeless in all vertical planes.

Nodal structure in basal plane is obtained using reflection operation $\sigma_z = IC_{2z}$, and taking into account that C_{2z} permutes basis functions and I results in factor -1 for triplet functions we have:

$$IC_{2z}\Psi(A_{1u}) = IC_{2z}\left(\psi^t_{k_1}e^{im\theta} - \psi^t_{\sigma_x k_1}e^{-im\theta} + \psi^t_{C_{2z}k_1}e^{im\theta} - \psi^t_{\sigma_y k_1}e^{-im\theta}\right) =$$
$$\left(-\psi^t_{C_{2z}k_1}e^{im\theta} + \psi^t_{\sigma_y k_1}e^{-im\theta} - \psi^t_{k_1}e^{im\theta} + \psi^t_{\sigma_x k_1}e^{-im\theta}\right) = -\Psi(A_{1u})$$
$$IC_{2z}\Psi(B_{1u}) = IC_{2z}\left(\psi^t_{k_1}e^{i2\theta} + \psi^t_{\sigma_x k_1}e^{-i2\theta} + \psi^t_{C_{2z}k_1}e^{i2\theta} + \psi^t_{\sigma_y k_1}e^{-i2\theta} =\right) =$$
$$\left(-\psi^t_{C_{2z}k_1}e^{i2\theta} - \psi^t_{\sigma_y k_1}e^{-i2\theta} - \psi^t_{k_1}e^{i2\theta} - \psi^t_{\sigma_x k_1}e^{-i2\theta}\right) = -\Psi(B_{1u})$$
$$IC_{2z}\Psi(B_{2u}) = IC_{2z}\left(\psi^t_{k_1}e^{i\theta} + \psi^t_{\sigma_x k_1}e^{-i\theta} - \psi^t_{C_{2z}k_1}e^{i\theta} - \psi^t_{\sigma_y k_1}e^{-i\theta}\right) =$$
$$\left(-\psi^t_{C_{2z}k_1}e^{i\theta} - \psi^t_{\sigma_y k_1}e^{-i\theta} + \psi^t_{k_1}e^{i\theta} + \psi^t_{\sigma_x k_1}e^{-i\theta}\right) = \Psi(B_{2u})$$
$$IC_{2z}\Psi(B_{3u}) = IC_{2z}\left(\psi^t_{k_1}e^{i\theta} - \psi^t_{\sigma_x k_1}e^{-i\theta} - \psi^t_{C_{2z}k_1}e^{i\theta} + \psi^t_{\sigma_y k_1}e^{-i\theta}\right) =$$
$$\left(-\psi^t_{C_{2z}k_1}e^{i\theta} + \psi^t_{\sigma_y k_1}e^{-i\theta} + \psi^t_{k_1}e^{i\theta} - \psi^t_{\sigma_x k_1}e^{-i\theta}\right) = \Psi(B_{3u})$$

Thus A_{1u} and B_{1u} have nodes in the basal plane, and B_{2u} and B_{3u} are nodeless in the basal plane. The nodal structure of pairs' functions of odd IRs of D_{2h} with phase winding, obtained by direct analysis of basis function, is presented in Table 4.29. This structure is the same as that in Table 4.25, obtained with application of Mackey and Frobenius theorems.

Since reflections in vertical planes change the phase winding direction, the extension of phase winding from basis domain of a BZ by point group elements results in zero total phase winding in the entire BZ.

Now it will be shown that to obtain phase winding $2\pi m$ the reflections should be performed with time reversal. Let us take up two pair functions $\psi_{k\alpha}$ and $\psi_{k\beta}$ with azimuthal angles α and $\beta > \alpha$, whose phase factors are $\exp(im\alpha)$ and $\exp(im\beta)$. Consider a reflection σ_y, which transforms these functions into positions $-\alpha$ and $-\beta$, but not affecting the phase factors. Thus a counterclockwise rotation from $-\beta$ to $-\alpha$ corresponds to an increase of the angle but phase winding factor changing from larger value $\exp(im\beta)$ to a smaller value $\exp(im\alpha)$, i.e. corresponds to magnetic quantum number $-m$. Consider now the action of the element $\vartheta\sigma_y$. Since time reversal is antilinear, its action on the phase factors results in a complex conjugation, i.e. phase factors at $-\beta$ and $-\alpha$ are $\exp(-im\beta)$ and $\exp(-im\alpha)$. Thus a counterclockwise rotation from $-\beta$ to $-\alpha$ corresponds to a phase increase, i.e. the magnetic quantum number remains unchanged.

Now let's consider the magnetic group $m'm'm$, in which the elements C_{2x}, C_{2y}, σ_x and σ_y appear with time reversal ϑ. Making use of Herring criterion (4.11.2) it was obtained that corepresentations of magnetic group $m'm'm$ belong to type (a), i.e. for unitary subgroup they are just IRs (4.11.3) and for non-unitary left cosets they extended with signs + or − (4.11.4), which are used in Table 4.27 as superscripts.

Bearing in mind that the reflection with time reversal doesn't change the phase winding direction, one obtains phase structures for corepresentation A_u^+ and A_u^- shown in Fig. 4.19a, b. Both structures have total phase winding 4π, but they differ by nodal structures. In the case of A_u^+, the phase is continuous when passing through the coordinate axes (coordinate planes in three dimensions) and the structure is nodeless. The phase of A_u^- changes by π when passing through the axes, resulting in nodes, denoted by two bold red lines. Total winding in the structures of B_u^+ and B_u^- equals to 2π and nodal structure in vertical planes is similar to that of A_u^+ and A_u^-. Functions with $m = 2$ and $m = -1$ differ by nodal structure in the basal plane, namely, A_u^+ and A_u^- are nodal but B_u^+ and B_u^- are nodeless.

Example 4.1 Structure of two-electron states in magnetic group $4/mm'm'$ with phase winding $m = 1$ and $m = 2$.

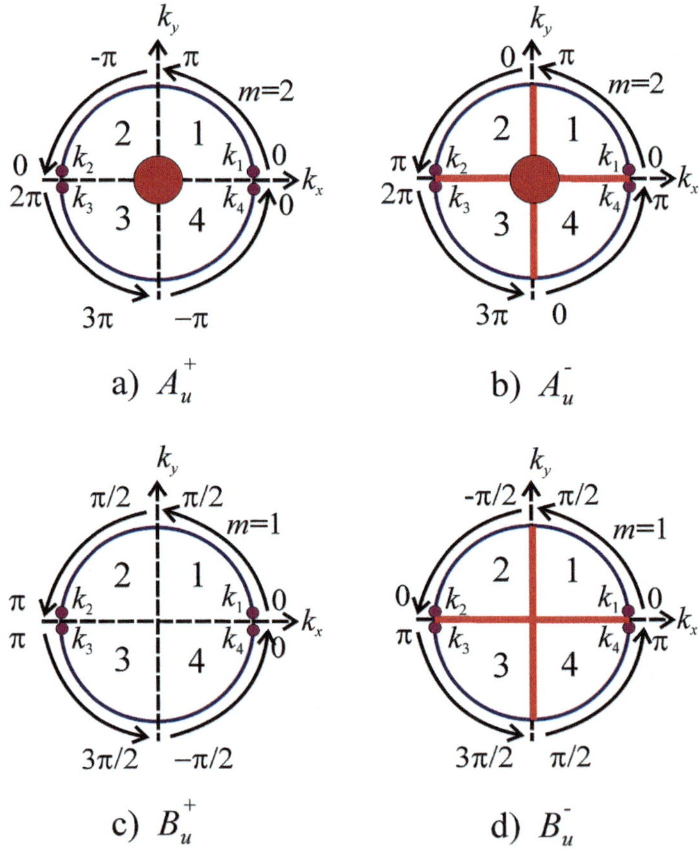

Fig. 4.19 Nodal structure and phase winding of SOP in $m'm'm$ symmetry for odd ICRs **a** A_u^+, **b** A_u^-, **c** B_u^+ and **d** B_u^-. Bold red lines denote vertical nodal planes and red circles denote nodal basal plane

Choosing antiunitary element $A = \vartheta\sigma_y$ as a representative of the antiunitary left coset, we represent the magnetic group $4/mm'm'$ in the form:

$$M = \mathsf{C}_{4h} + \vartheta\sigma_y\mathsf{C}_{4h}$$

It is immediately verified that, since spatial parts of anti-unitary elements B are reflections in vertical planes and rotations around horizontal axes by angle π, all squares of anti-unitary elements B^2 result in an identity element. Since the spin is integer, the square of time reversal ϑ^2 results in one. Thus the sum in Herring criterion can be easily estimated:

$$\sum_{B\in\vartheta\sigma_y\mathsf{C}_{4h}} \chi\left(B^2\right) = 8$$

Hence it follows that ICRs belong to type (a), and they are defined by formula (4.11.3) for unitary subgroup and by formula (4.11.4) for non-unitary left coset. ICRs of group $4/mm'm'$ are presented in Table 4.30 and they are labeled by IRs of unitary subgroup, with the subscript, corresponding to the sign in non-unitary subgroup. Let initial k_1 be close to k_x-axis and pair's function $\psi_1^{s(t)}$ be real and positive. In this case, $\psi_1^{s(t)}$ is unaffected by time reversal ϑ and the factors before the other functions are defined by characters of ICRs only. It is seen in Fig. 4.7 that when initial k_1 approaches k_x-axis, pairs of functions $(\psi_1^{s(t)}, \psi_{27}^{s(t)})$, $\left(\psi_{14}^{s(t)}, \psi_{37}^{s(t)}\right)$, $\left(\psi_4^{s(t)}, \psi_{26}^{s(t)}\right)$ and $\left(\psi_{15}^{s(t)}, \psi_{40}^{s(t)}\right)$ meet at the axes k_x, k_y, $-k_x$ and $-k_y$, respectively. It is seen in Table 4.30 that the characters for these pairs of elements for ICRs with superscript + are the same. Hence, it follows that the phases of functions in each one of these pairs are the same and there are no nodes in directions k_x, k_y, $-k_x$ and $-k_y$ (in planes (100) and (010) in three dimensions). Also the functions $(\psi_1^{s(t)}, \psi_{37}^{s(t)})$, $\left(\psi_{14}^{s(t)}, \psi_{26}^{s(t)}\right)$, $\left(\psi_4^{s(t)}, \psi_{40}^{s(t)}\right)$ and $\left(\psi_{15}^{s(t)}, \psi_{27}^{s(t)}\right)$ meet in diagonal planes with the same phases. Hence, it follows that the basis functions of ICRs with + superscript are nodeless in all planes and basis functions of ICRs with superscript − are nodal in all vertical planes. Since time reversal is connected with reflections in vertical planes, the phase winding is the same in all sectors. It can be easily obtained from

Table 4.30 ICRs of group $4/mm'm'$ and possible angular momentum projection $\bar{m}$

	h_1	h_{14}	h_4	h_{15}	ϑh_{27}	ϑh_{37}	ϑh_{26}	ϑh_{40}	$\bar{m}$
ICR	E	C_{4z}	C_{2z}	C_{4z}^3	$\vartheta\sigma_y$	$\vartheta\sigma_a$	$\vartheta\sigma_x$	$\vartheta\sigma_b$	
$A_g^{\pm}(A_u^{\pm})$	1	1	1	1	±1	±1	±1	±1	$0, \pm4$
$B_g^{\pm}\,(B_u^{\pm})$	1	-1	1	-1	±1	∓1	±1	∓1	±2
$E_{1g}^{\prime\pm}\,(E_{1u}^{\prime\pm})$	1	i	-1	$-i$	±1	$\pm i$	∓1	$\mp i$	$+1$
$E_{2g}^{\prime\pm}\,(E_{1u}^{\prime\pm})$	1	$-i$	-1	i	±1	$\mp i$	∓1	$\pm i$	-1

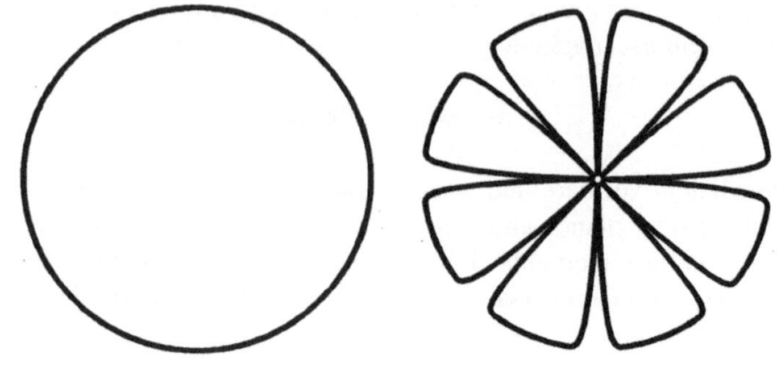

Fig. 4.20 SOP structures for ICRs induced by IRs $A^{\pm}_{g(u)}$, $B^{\pm}_{g(u)}$, $E'^{\pm}_{1g(u)}$ and $E'^{\pm}_{2g(u)}$ of unitary subgroup C_{4h} in magnetic symmetry $4/mm'm'$. **a** + sign for non-unitary elements (as in Table 4.30), **b** – sign for non-unitary elements

this property that when the phase winding in the first sector is projected on the other sectors the total phase winding is $\pm 4\pi$ for $B^{\pm}_{g}$ $(B^{\pm}_{u})$, 2π for $E'^{\pm}_{1g}$ $(E'^{\pm}_{1u})$ and -2π for $E'^{\pm}_{2g}$ $(E'^{\pm}_{2u})$. For $A^{\pm}_{g}$ $(A^{\pm}_{u})$ phase winding may be 0 or 8π. The structure of SOP obtained with account for the interference of complex phase of basis functions of ICRs listed in Table 4.30 is shown in Fig. 4.20. The of nodes in vertical planes are the same for all winding numbers. The functions belonging to ICRs in Table 4.30 with + sign are nodeless (see Fig. 4.20a) and the functions belonging to IRCs in Table 4.30 with – sign are nodal (see Fig. 4.20b). In the latter case, phase difference at all vertical planes equals to π, but the phase winding direction not changes and total phase winding also not changes with respect to the former case. Note that the nodal structure of basis functions of ICRs in basal plane is the same as that of corresponding IRs of D_{4h} group.

4.13 Possible SOP Structures in D_{6h} and $6/mm'm'$ Symmetry

Group D_{6h} is the largest axial crystallographic point group and in this section we consider structures of SOP corresponding to its two-dimensional IRs and also the structure of SOP in magnetic group $6/m'm'm$ in which reflections in vertical planes are connected with time reversal. We consider nodal structure and phase winding of odd two-dimensional IRs. Nodal structure of SOP corresponding to even IRs is very similar and differs only in the basal plane.

It was shown in previous sections that in order to identify magnetic quantum numbers corresponding to IRs of D_{nh} group it is sufficient to consider its subgroup

Table 4.31 IRs D^{κ} of group C_{6h}, their magnetic quantum numbers $\bar{m}$, decomposition of induced representations, and nodes in basal plane (001)

D^{κ}	E	C_{6z}^1	C_{6z}^2	C_{6z}^3	C_{6z}^4	C_{6z}^5	m	$D^{\kappa} \uparrow D$	Nodes (001)
A_u	1	1	1	1	1	1	0	$A_{1u} + A_{2u}$	Yes
B_u	1	-1	1	-1	1	-1	± 3	$B_{1u} + B_{2u}$	no
E'_{1u}	1	ω	ω^2	-1	ω^4	ω^5	1	$\tilde{E}_{1u}$	No
E''_{1u}	1	ω^5	ω^4	-1	ω^2	ω	-1	$\tilde{E}^*_{1u}$	No
E'_{2u}	1	ω^2	ω^4	1	ω^2	ω^4	2	$\tilde{E}_{2u}$	Yes
E''_{2u}	1	ω^4	ω^2	1	ω^4	ω^2	-2	$\tilde{E}^*_{2u}$	Yes

Table 4.32 Some matrices of IRs $\tilde{E}_{1u}$ and $\tilde{E}_{2u}$ of D_{6h} group in complex form. Matrices for the other elements of group C_{6v} may be obtained multiplying matrices for rotations by the matrix for σ_1 (reflection in (010) plane)

	C_{6z}^1 (C_1)	C_{6z}^2 (C_2)	C_{6z}^3 (C_3)	C_{6z}^4 (C_4)	C_{6z}^5 (C_3)	σ_1
$\tilde{E}_{1g(u)}$	$\begin{pmatrix} \omega & 0 \\ 0 & \omega^5 \end{pmatrix}$	$\begin{pmatrix} \omega^2 & 0 \\ 0 & \omega^4 \end{pmatrix}$	$\begin{pmatrix} -1 & 0 \\ 0 & -1 \end{pmatrix}$	$\begin{pmatrix} \omega_4 & 0 \\ 0 & \omega_2 \end{pmatrix}$	$\begin{pmatrix} \omega^5 & 0 \\ 0 & \omega \end{pmatrix}$	$\begin{pmatrix} 0 & 1 \\ 1 & 0 \end{pmatrix}$
$\tilde{E}_{2g(u)}$	$\begin{pmatrix} \omega^2 & 0 \\ 0 & \omega^4 \end{pmatrix}$	$\begin{pmatrix} \omega^4 & 0 \\ 0 & \omega^2 \end{pmatrix}$	$\begin{pmatrix} 1 & 0 \\ 0 & 1 \end{pmatrix}$	$\begin{pmatrix} \omega^2 & 0 \\ 0 & \omega^4 \end{pmatrix}$	$\begin{pmatrix} \omega^4 & 0 \\ 0 & \omega^2 \end{pmatrix}$	$\begin{pmatrix} 0 & 1 \\ 1 & 0 \end{pmatrix}$

C_{nh}. IRs of C_{6h} group for rotations, their magnetic quantum numbers, results of their induction into D_{6h} and nodes in basal plane are shown in Table 4.31. It is immediately verified that the phase factors for finite rotations of IRs A_u, B_u, E'_{1u}, E''_{1u}, E'_{2u} and E''_{2u} correspond to angular momentum projections $\bar{m} = 0, \pm 3, 1, -1, 2$ and -2. The representations $\tilde{E}_{1u}$ and $\tilde{E}_{2u}$ of group D_{6h} obtained by the induction of E'_{1u} and E'_{2u} are presented in Table 4.32. The induction of E''_{1u} and E''_{2u} results in complex conjugate matrices, denoted as $\tilde{E}^*_{1u}$ and $\tilde{E}^*_{2u}$ which are equivalent to $\tilde{E}_{1u}$ and $\tilde{E}_{2u}$, since all characters are real. Matrices for the reflection elements are obtained multiplying by matrix for σ_1 (reflection in (010) plane). These matrices are IRs of elements belonging to C_{6v} subgroup of D_{6h}. The remaining matrices of the left coset $I\mathsf{C}_{6v}$ may be obtained using parity of IRs. It can be verified that multiplication of $\tilde{E}_{1u}$ and $\tilde{E}_{2u}$ by IR A_{2g} results in also $\tilde{E}_{1u}$ and $\tilde{E}_{2u}$. This multiplication is equivalent to changing sign of the matrix for σ_1 in Table 4.32. This matrices are denoted by superscripts $+$ and $-$. This superscript is an additional (internal) quantum number, which doesn't change the character of IR, but changes the structure of the wavefunction. Due to the choice of the same matrices σ_1 for even and odd IRs they have identical nodal structures in vertical planes for the same superscripts. For example, nodal structures in vertical planes of $\tilde{E}^+_{1u}$ and $\tilde{E}^+_{1g}$ are the same. However, even and odd IRs have different nodal structure in a basal plane.

The basis set of a Cooper pair for D_{6h} symmetry is shown in Fig. 4.21. The basis of a pair is obtained by the action on the function ψ_1 (see formula (4.7.1)) in the

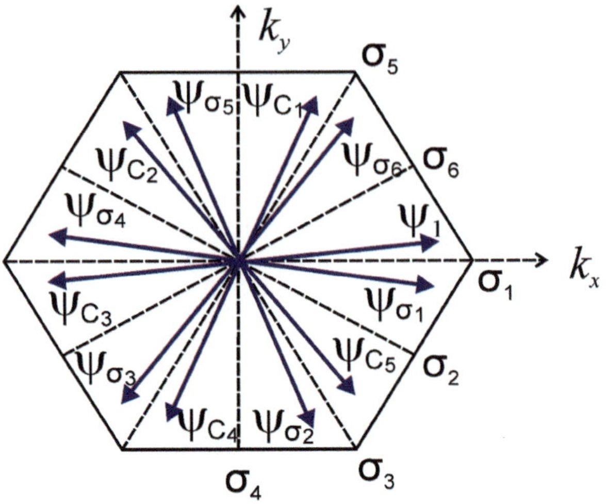

Fig. 4.21 Basis set of a Cooper pair in D_{6h} symmetry, subscripts of rotational elements C_i correspond to subscripts in Tables 4.32 and 4.33

Table 4.33 Basis pair functions of IRs $\tilde{E}_{1u}^{\pm}$ and $\tilde{E}_{2u}^{\pm}$ of D_{6h} group in chiral form and $\bar{m}-$values. See Fig. 4.21 for notations

IR, row	Function	$\bar{m}$
$\Psi_1(\tilde{E}_{1u}^{\pm})$	$\psi_1^t + \omega\psi_{C_1}^t + \omega^2\psi_{C_2}^t - \psi_{C_3}^t + \omega^4\psi_{C_4}^t + \omega^5\psi_{C_5}^t$	1
$\Psi_2(\tilde{E}_{1u}^{\pm})$	$\pm\psi_{\sigma_1}^t \pm \omega^5\psi_{\sigma_6}^t \pm \omega^4\psi_{\sigma_5}^t \mp \psi_{\sigma_4}^t \pm \omega^2\psi_{\sigma_3}^t \pm \omega\psi_{\sigma_2}^t$	-1
$\Psi_1(\tilde{E}_{2u}^{\pm})$	$\psi_1^t + \omega^2\psi_{C_1}^t + \omega^4\psi_{C_2}^t + \psi_{C_3}^t + \omega^2\psi_{C_4}^t + \omega^4\psi_{C_5}^t$	2
$\Psi_2(\tilde{E}_{2u}^{\pm})$	$\pm\psi_{\sigma_1}^t \pm \omega^4\psi_{\sigma_6}^t \pm \omega^2\psi_{\sigma_5}^t \pm \psi_{\sigma_4}^t \pm \omega^4\psi_{\sigma_3}^t \pm \omega^2\psi_{\sigma_2}^t$	-2

representation domain with the rotations C_i about $z-$axis and reflections in vertical planes σ_i shown in Fig. 4.21. These elements together with the identity element form a group C_{6v} consisting of 12 elements. When the initial k runs over the basis domain of a BZ, the other $k-$vectors run over all other domains of a BZ.

The basis functions for two-dimensional IRs $\tilde{E}_{1u}^{\pm}$ and $\tilde{E}_{2u}^{\pm}$ obtained by projection operator (4.7.2) are presented in Table 4.33. It is seen in Table 4.33 that the basis functions of the first row connected by rotations appear with phase factor, corresponding to $\bar{m} = 1$ for and $\bar{m} = 2$ for $\tilde{E}_{1u}^{\pm}$ and $\tilde{E}_{2u}^{\pm}$, respectively. Basis functions of the second rows of $\tilde{E}_{1u}^{\pm}$ and $\tilde{E}_{2u}^{\pm}$ correspond to $\bar{m} = -1$ and $\bar{m} = -2$, respectively. To obtain a pair momentum in a plane normal to k_z- we equate $\bar{m}-$values for discrete rotations to $m-$values for phase winding in the first sector. In the sectors, connected by rotations with the first sector angular momentum projections are: $m = 1$ or $m = 2$. Since the reflection changes the sign of the phase winding in the sectors connected with the first one with reflections angular momentum projections are: $m = -1$ or $m = -2$. Thus total phase winding equals to zero.

However, there is one more quantum number—the sign of matrix of left coset representative σ_1. When the initial wavevector approaches a plane (010) two mirror counterparts merge in the case of $\tilde{E}_{1u}^{+}$ and $\tilde{E}_{2u}^{+}$ or cancel in the case of $\tilde{E}_{1u}^{-}$ and $\tilde{E}_{2u}^{-}$.

On the symmetry planes interference takes place of real and imaginary parts of pair functions, located on opposite sides of the mirror plane. In order to envisage such interference, the wavefunctions of a pairs in a finite number of points were represented by real and imaginary Gaussians with coefficients equal to the values of real and imaginary parts of the pair's function. Within the sectors the phases at neighboring points almost coincide and constructive interference takes place. At the boundaries of the sectors, both constructive and destructive interferences of the wavefunctions of pairs are possible. When squaring modulus of the complex function at every point we obtain the structure of SOP in the plane normal to k_z-axis. The results for $\tilde{E}_{1u}^{+}$ and $\tilde{E}_{1u}^{-}$ are shown in Fig. 4.22a, b, respectively. Nodes of $\tilde{E}_{1u}^{+}$ at k_y-axis in Fig. 4.22a appear since real parts are zero, and imaginary parts have opposite signs. The node of $\tilde{E}_{1u}^{-}$ at k_x-axis appears since in this case reflection σ_1 is connected with the change of the sign. There can also exist constructive interference of real parts and destructive interference of imaginary parts, and vice versa, resulting in dips. At other vertical mirror planes in Fig. 4.22a, b there are dips (not nodes).

The structure of $\tilde{E}_{2u}^{+}$ shown in Fig. 4.23a is nodeless with similar dips only. The structure of $\tilde{E}_{2u}^{-}$ in Fig. 4.23b has nodes in k_x- and k_y-axes (planes (010) and (100) in three dimensions) and dips in other planes.

Similar to the case of group D_{2h}, considered in a previous section we consider for D_{6h} point group a magnetic group $6/mm'm'$ with unitary subgroup C_{6h} in which time reversal is connected with reflections in vertical planes. When applying Herring's criterion, we obtain that all corepresentations of $6/mm'm'$ belong to type (a), i.e. they are IRs of C_{6h} extended on antiunitary elements with the sign $+$ or $-$. These corepresentations $\Delta^{\pm}(D^k)$ which all are one-dimensional are denoted by the label

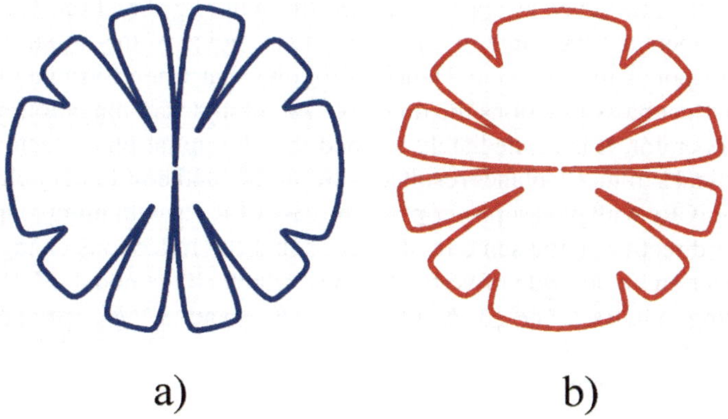

Fig. 4.22 Nodal structure of squared two-electron wavefunction with phase winding for complex IR of D_{6h}. **a** $\tilde{E}_{1u}^{+}$ **b** $\tilde{E}_{1u}^{-}$

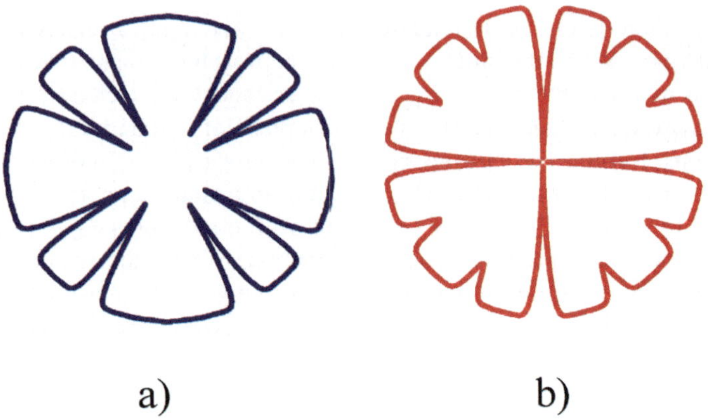

Fig. 4.23 Nodal structure of squared two-electron wavefunction with phase winding for complex IR of D_{6h}. **a** $\tilde{E}^{+}_{2u}$ **b** $\tilde{E}^{-}_{2u}$

Table 4.34 Corepresentations $\Delta^{\pm}$ of magnetic group $6/mm'm'$, induced by two complex IRs E'_{1u} and E'_{2u} of C_{6h}. Shorthand notations of Fig. 4.21 for the group elements are used

	E	C_1	C_2	C_3	C_4	C_5	$\vartheta\sigma_1$	$\vartheta\sigma_2$	$\vartheta\sigma_3$	$\vartheta\sigma_4$	$\vartheta\sigma_5$	$\vartheta\sigma_6$
$\Delta^{\pm}(E'_{1u})$	1	ω	ω^2	-1	ω^4	ω^5	± 1	$\pm\omega^5$	$\pm\omega^4$	∓ 1	$\pm\omega^2$	$\pm\omega$
$\Delta^{\pm}(E'_{2u})$	1	ω^2	ω^4	1	ω^2	ω^4	± 1	$\pm\omega^4$	$\pm\omega^2$	± 1	$\pm\omega^4$	$\pm\omega^2$

of IRs of its unitary subgroup D^k and corresponding superscripts are presented in Table 4.34.

It follows form Table 4.34 that transformations of the initial function ψ_1 by unitary elements (pure rotations) of $\Delta^{\pm}(E'_{1u})$ and $\Delta^{\pm}(E'_{2u})$ correspond to magnetic quantum numbers 1 and 2, respectively. Since, as it was shown in previous section, reflection with time reversal doesn't change the winding direction, the magnetic quantum numbers are the same in all sectors. Our choice of the initial ψ_1 close to k_x-axis corresponds to real value. Hence, it follows that when formula (4.11.10) is used to construct basis functions the time reversal (complex conjugation) is reduced to identity operator. It is immediately verified that the initial phase factors on both sides of vertical symmetry planes resulted in by unitary and non-unitary elements are the same for ICRs with plus superscript. In the case of ICRs with minus superscripts, the winding direction is the same in all sectors, but the reflections change the sign of a pair function on one side of vertical planes, which all are nodal. SOP structures corresponding to ICRs $\Delta^{+}(E'_{1u})$, $\Delta^{+}(E'_{2u})$, $\Delta^{-}(E'_{1u})$ and $\Delta^{-}(E'_{2u})$ are presented in Fig. 4.24. As it seen in Fig. 4.24a, the SOPs for $\Delta^{+}(E'_{1u})$ and $\Delta^{+}(E'_{2u})$ are nodeless in all vertical planes, but the SOPs $\Delta^{-}(E'_{1u})$ and $\Delta^{-}(E'_{2u})$ in Fig. 4.24b have nodes in all vertical planes. These vertical nodes depend on additional quantum number, i.e. plus or minus sign on non-unitary left coset representative σ_1 and are topologically unstable. However, in basal plane, the nodes of $\Delta^{+}(E'_{2u})$ and $\Delta^{-}(E'_{2u})$ and also

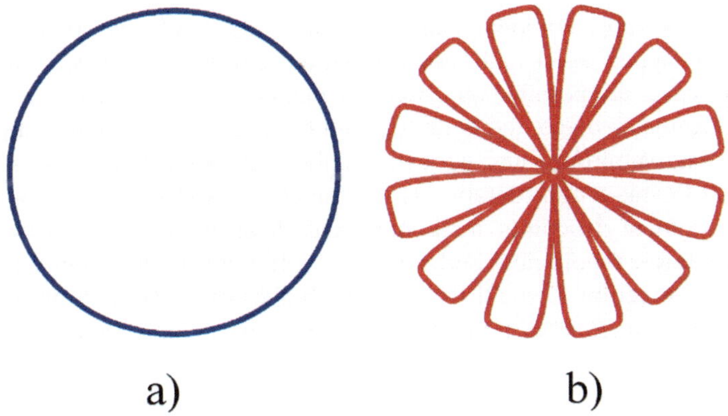

Fig. 4.24 Nodal structure of squared two-electron wavefunction with phase winding for ICR of $6/mm'm'$. **a** $\Delta^+(E'_{1u})$ and $\Delta^+(E'_{2u})$ **b** $\Delta^-(E'_{1u})$ and $\Delta^-(E'_{2u})$

the nodeless structure of $\Delta^+(E'_{1u})$ and $\Delta^-(E'_{1u})$ are topologically stable and are the same as nodes of corresponding IRs of C_{6h} and D_{6h} (see Table 4.31).

Let us consider some nontrivial relations for the symmetry of electron pairs in singular points of the BZ for the space groups D^1_{6h} and D^4_{6h}, which is shown in Fig. 4.25. The decompositions of antisymmetrized squares of IRs for symmetry directions and planes are presented in Table 4.35. Nodal planes can be divided into group-theoretical and topological ones. For one-dimensional IRs, all planes where these IRs don't appear in the decomposition of antisymmetrized (or symmetrized) square

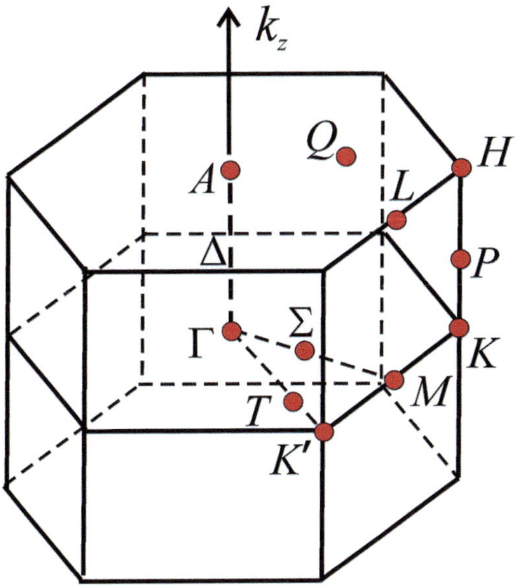

Fig. 4.25 BZ for the space groups D^1_{6h} and D^4_{6h}

are nodal in a group-theoretical sense. In general points of a BZ two-dimensional odd (even) IRs appear twice in the decomposition of antisymmetrized (symmetrized) square. Two cases are possible on symmetry planes. If two-dimensional IR appears once on a symmetry plane, the nodes depend on additional quantum numbers (IRs of intermediate subgroup) and nodes are called topological. In group-theoretical nodal planes IR under consideration is completely forbidden. For group-theoretical nodal planes, nodal structures in all equivalent planes are the same. However, as it was shown above, topological nodes may be different in equivalent planes. In all planes and in direction Σ small IRs are one-dimensional and possible symmetries of two-electron functions don't depend on the index of small IR (see Table 4.35). For two-dimensional IRs E_1 and E_2 in the direction Δ even IR A_{1g} appears in the decomposition of antisymmetrized square. This means that if one-electron IR is two-dimensional, triplet pairs with even spatial part of the wavefunctions are possible. Two-dimensional IRs E_{1u} and E_{2u} appear twice in a general point of a BZ, but ones in vertical planes. Thus both nodal and nodeless in vertical plane functions of E_{1u} and E_{2u} symmetry are possible. Nodes of two-dimensional IRs in vertical planes are topologically unstable, i.e. they depend on additional quantum numbers. In basal plane E_{2u} is completely forbidden and the nodes are group theoretical.

Also an interesting case appears for double-valued IRs in the direction Δ. In direction Δ little group is C_{6v} and IR of the space group is obtained by the induction of double-valued IRs E'_1, E'_2 or E'_3 from C_{6v} into D_{6h} (see Table 4.36). Decomposition of antisymmetrized squares results in for the Cooper terms:

$$\{E'_1 \uparrow D_{6h} \times E'_1 \uparrow D_{6h}\} = A_{1g} + A_{1u} + E_{1u}$$

$$\{E'_3 \uparrow D_{6h} \times E'_3 \uparrow D_{6h}\} = A_{1g} + A_{1u} + B_{1u} + B_{2u}$$

The results for E'_1 and E'_3 are different. Since in C_{6v} symmetry E'_1 corresponds to $j = 1/2$ and $j = 3/2$ decomposes into E'_1 and E'_3 this dependence of possible Cooper pair wavefunction on the small IR is called also j–dependence.

Table 4.35 Decomposition of antisymmetrized squares of IRs for symmetry planes and directions in the BZ for groups D^1_{6h} and D^4_{6h} (see Fig. 4.25 for notations of BZ points)

Planes	D^K	Antisymmetrized square
ΓML	$A, A\prime$	$A_{2u} + B_{2u} + E_{1u} + E_{2u}$
ΓKH	$A, A\prime$	$A_{2u} + B_{1u} + E_{1u} + E_{2u}$
ΓKM	$A, A\prime$	$B_{1u} + B_{2u} + 2E_{1u}$
Directions		
Σ	A_1, A_2, B_1, B_2	$B_{2u} + E_{1u}$
T	A_1, A_2, B_1, B_2	$B_{1u} + E_{1u}$
Δ	A_1, A_2, B_1, B_2	A_{2u}
Δ	E_1, E_2	$A_{1u} + E_{2u} + A_{1g}$

Example 4.2 Obtain possible Cooper pair wavefunction for direction HK in the BZ of groups D_{6h}^1 and D_{6h}^4 (see Fig. 4.25).

It is clear from Fig. 4.25 that rotations C_3 and C_3^2 move a point in HK-direction, to points connected by reciprocal lattice vectors. Thus the symmetry group H of the wavevector in HK direction is extended from C_s to C_{3v}. When applying Herring criterion for double-valued IR A_1' of C_{3v} we obtain

$$\nu\left(\theta^2\right)\sum_{H}\hat{\chi}\left(h^2\right)\,\omega_S\left(h,h,\right)=(-1)(1+1+1-1-1-1)=0 \tag{4.13.1}$$

The result is the same for A_2'. Thus due to time-reversal symmetry IRs A_1' and A_2' form one corepresentation of type (c), whose character is also presented in Table 4.37. Similar sum for E' is written as

$$\nu\left(\theta^2\right)\sum_{H}\hat{\chi}\left(h^2\right)\,\omega_S\left(h,h,\right)=(-1)\left(2-1-1-2-2-2\right)=6 \tag{4.13.2}$$

Since this sum equals to $\left|\bar{G}\right|$ there is no additional degeneration due to time-reversal symmetry for IR E'.

Data for calculation antisymmetrized squares of double-valued IRs are given in Table 4.37. The second row of the table shows the squared elements h^2.

When inducing representations of antisymmetrized squares from D_{3d} into D_{6h}, we obtain

$$P^{E'-}\uparrow D_{6h}=A_{1g}+A_{1u}+B_{2g}+B_{2u}+E_{1u}+E_{2u}$$

$$P^{\Delta(A_{12}')-}\uparrow D_{6h}=A_{1g}+2A_{1u}+A_{2u}+B_{1u}+B_{2g}+2B_{2u}$$

Table 4.36 Data for constructing antisymmetrized squares of double-valued IRs in the direction Δ in the BZ for groups D_{6h}^1 and D_{6h}^4 (see Fig. 4.25 for notations) ($H=\mathrm{C}_{6v}$, $\tilde{M}=D_{6h}$) The asterisk in g^2 line denotes the case when the product $(amam)$ in Mackey theorem results in angle more than 2π and a factor -1 is added

g	E	C_{6z}	C_{6z}^2	C_{6z}^3	C_{6z}^4	C_{6z}^5	$6\sigma_i$	I	S_6	S_3	σz	S_3^2	S_6^5	$6C_{2i}$
h	h_1	h_2	h_3	h_4	h_5	h_6		h_{13}	h_{14}	h_{15}	h_{16}	h_{17}		
g^2								E	C_{6z}^2	C_{6z}^4	E^*	C_{6z}^{2*}	C_{6z}^{4*}	E^*
E_1'	2	$\sqrt{3}$	1	0	-1	$-\sqrt{3}$	0							
E_2'	2	$-\sqrt{3}$	1	0	-1	$\sqrt{3}$	0							
E_3'	2	0	-2	0	2	0	0							
$\{E_1'^2\}$	4	3	1	0	1	3	0	-2	-1	1	2	1	-1	2
$\{E_3'^2\}$	4	0	4	0	4	0	0	-2	2	-2	2	-2	2	2

Table 4.37 Data for constructing antisymmetrized squares of double-valued IRs in the direction HK in the BZ for groups D^1_{6h} and D^4_{6h} (see Fig. 4.25) ($H = C_{4v}$, $\tilde{M} = D_{3d}$). The asterisk denotes the case when the product $(amam)$ in Mackey theorem results in angle more than 2π and a factor -1 is added

h	E	C_3	C_3^2	σ_1	σ_3	σ_5	I	IC_3	IC_3^2	$I\sigma_1$	$I\sigma_3$	$I\sigma_5$
h^2	E	C_3^2	C_3^*	E^*	E^*	E^*	E	C_3^2	C_3^*	E^*	E^*	E^*
A'_1	1	-1	1	i	$-i$	i						
A'_2	1	-1	1	$-i$	i	$-i$						
E'	2	1	-1	0	0	0						
$\Delta(A'_{1,2})$	2	-2	2	0	0	0						
$P^{E'-}$	4	1	1	0	0	0	-2	1	1	2	2	2
$P^{\Delta(A'_{1,2})-}$	4	4	4	0	0	0	-2	-2	-2	2	2	2

Thus possible symmetries of Cooper pairs are different in these two cases. In the case of small IR E' pairs of E_{1u} and E_{2u} symmetry are possible in the symmetry direction HK, but for $\Delta(A'_{1,2})$ pairs E_{1u} and E_{2u} are forbidden in this direction.

4.14 Additional Quantum Numbers on Symmetry Lines

We consider as an example T_h^5 $(Im3)-$ a symmetry group of triplet superconductor $PsOs_4Sb_{12}$, for which point nodes were obtained in directions (100), (010) and (001) [97]. In the directions k_x, k_y and k_z, the wavevector groups are isomorphic to C_{2v} and the group $\tilde{M}$ in Mackey theorem is D_{2h}, which is invariant subgroup of T_h. Characters of IRs of T_h and of D_{2h} are presented in Table 4.38.

Table 4.38 IRs of group T_h and of its invariant subgroup D_{2h}. $\varepsilon = \exp(i2\pi/3)$

T_h	E	C_{2x}	C_{2y}	C_{2z}	$4C_3$	$4C_3^2$	I	σ_x	σ_y	σ_z	$4S_6^5$	$4S_6$
$A_{1g(u)}$	1	1	1	1	1	1	±1	±1	±1	±1	±1	±1
$E'_{g(u)}$	1	1	1	1	ε	ε^2	±1	±1	±1	±1	$\pm\varepsilon$	$\pm\varepsilon^2$
$E''_{g(u)}$	1	1	1	1	ε^2	ε	±1	±1	±1	±1	$\pm\varepsilon^2$	$\pm\varepsilon$
$T_{g(u)}$	3	-1	-1	-1	0	0	±3	∓1	∓1	∓1	0	0
D_{2h}												
$A_{g(u)}$	1	1	1	1			±1	±1	±1	±1		
$B_{1g(u)}$	1	-1	-1	1			±1	∓1	∓1	±1		
$B_{2g(u)}$	1	-1	1	-1			±1	∓1	±1	∓1		
$B_{3g(u)}$	1	1	-1	-1			±1	±1	∓1	∓1		

Let k be a wavevector at general point in the representation domain of a BZ of space group T_h^5 $(Im3)$ shown in Fig. 4.26. In a general point of a BZ spatial part of the wavefunction of a triplet pair belongs to IR A_u of C_i group. The induction into the whole group results in

$$A_{u(\mathsf{C}_i)} \uparrow T_h = A_u + E'_u + E''_u + 3T_u \tag{4.14.1}$$

Since a three-dimensional IR T_h appears three times in a general point of a BZ, additional quantum numbers are required to label repeating IRs. To obtain additional quantum numbers for T_u, we induce firstly IR A_u of C_i into intermediate subgroup D_{2h}:

$$A_{u(\mathsf{C}_i)} \uparrow D_{2h} = A_u + B_{1u} + B_{2u} + B_{3u} \tag{4.14.2a}$$

Further induction of IRs in r.h.s. of (4.14.7) into T_h results in

$$A_{u(\mathsf{C}_i)} \uparrow T_h = A_{1u} + E'_u + E''_u \tag{4.14.3}$$

$$B_{1u(\mathsf{C}_i)} \uparrow T_h = T_u \tag{4.14.4}$$

$$B_{2u(\mathsf{C}_i)} \uparrow T_h = T_u \tag{4.14.5}$$

$$B_{3u(\mathsf{C}_i)} \uparrow T_h = T_u \tag{4.14.6}$$

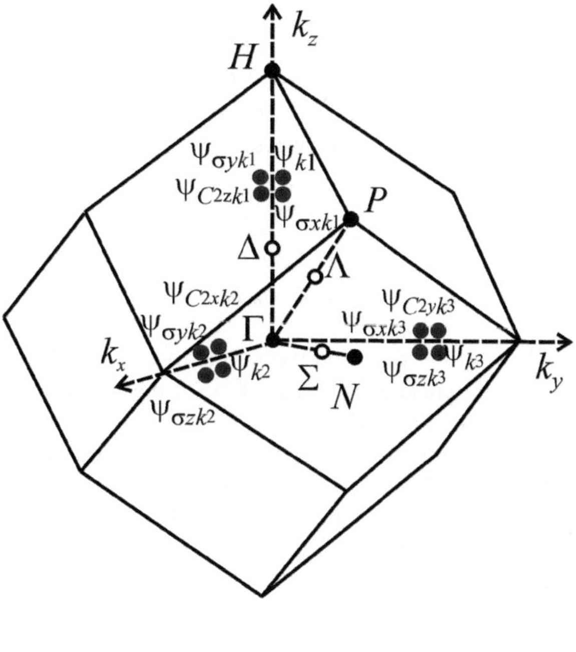

Fig. 4.26 Brillouin zone for T_h group. The pair is denoted by purple circles

Thus repeating IRs T_u may be labeled by the IRs B_{1u}, B_{2u} and B_{3u} of intermediate subgroup D_{2h} as $T_u^{B_{1u}}$, $T_u^{B_{2u}}$ and $T_u^{B_{3u}}$. In order to find out the physical meaning of this additional quantum number, we will analyze corresponding pairs' functions near the coordinate axes.

For k in $\Delta-$direction (see Fig. 4.26), the wavevector group H is a group C_{2v}. Making use of Mackey theorem, we obtain that the characters P_I^- for the elements C_{2v} equal to $+1$ and characters P_I^- for the elements IC_{2v} equal to -1. It is seen from Table 4.38 that it corresponds to IR to B_{1u} of D_{2h}. Note that there are three C_{2v} subgroups of T_h with rotational axes C_{2z}, C_{2x} and C_{2y}, but the extension of these group by inversion results in just one group D_{2h}. In the $\Delta-$direction we consider C_{2v} group with rotation C_{2z}. Consider now vector k in the vicinity of $\Delta-$direction and choose a pair basis set as follows. The first four elements of the basis set, namely, ψ_k, $\psi_{C_{2z}k}$, $\psi_{\sigma_x k}$ and $\psi_{\sigma_y k}$ are obtained by the action of the elements of C_{2v} on the initial $k-$vector. The other eight basis vectors are obtained by the actions of rotations C_3 and C_3^2 about $\Lambda-$direction on these basis sets. This pair basis set is denoted in Fig. 4.26 by purple circles. The basis sets of IRs B_{1u}, B_{2u} and B_{3u} in r.h.s. of (4.14.7) can be easily constructed by projection operators technique acting by the elements of C_{2v} on initial pair function ψ_k:

$$\Psi(B_{1u}) = \Psi_3(T_u^{B_{1u}}) = \psi_{k_1} + \psi_{C_{2z}k_1} + \psi_{\sigma_x k_1} + \psi_{\sigma_y k_1} \tag{4.14.7}$$

$$\Psi(B_{2u}) = \Psi_3(T_u^{B_{2u}}) = \psi_{k_1} - \psi_{C_{2z}k_1} + \psi_{\sigma_x k_1} - \psi_{\sigma_y k_1} \tag{4.14.8}$$

$$\Psi(B_{3u}) = \Psi_3(T_u^{B_{3u}}) = \psi_{k_1} - \psi_{C_{2z}k_1} - \psi_{\sigma_x k_1} + \psi_{\sigma_y k_1} \tag{4.14.9}$$

It is seen from formula (4.14.7), that basis function $\Psi(B_{1u})$ is invariant under the actions of the elements of C_{2v} group. Hence, it follows that when the initial k_1-vector approaches k_z-axis all basis functions $\Psi(B_{1u})$ merge and there is no node in k_z-direction. However for $\Psi(B_{2u})$ we have:

$$C_{2z}\left(\psi_{k1} - \psi_{C_{2z}k_1} + \psi_{\sigma_x k_1} - \psi_{\sigma_y k_1}\right) = \psi_{C_{2z}k_1} - \psi_{k_1} + \psi_{\sigma_y k_1} - \psi_{\sigma_x k_1}$$
$$\sigma_x\left(\psi_{k_1} - \psi_{C_{2z}k_1} + \psi_{\sigma_x k_1} - \psi_{\sigma_y k_1}\right) = \psi_{\sigma_x k_1} - \psi_{\sigma_y k_1} + \psi_{k_1} - \psi_{C_{2z}k_1}$$
$$\sigma_y\left(\psi_{k1} - \psi_{C_{2z}k1} + \psi_{\sigma_x k1} - \psi_{\sigma_y k1}\right) = \psi_{\sigma_y k1} - \psi_{\sigma_x k1} + \psi_{C_{2z}k1} - \psi_{k1}$$

Thus in k_z-direction basis functions of $\Psi(B_{2u})$ with the same sign merge and that with opposite signs cancel, resulting in a node. Similar result can be obtained for the function $\Psi(B_{3u})$.

Now it remains to prove that these properties are preserved on the axes k_x and k_y. The induced basis sets for T_h are presented in Table 4.39. The basis set for IR T_u, for the initial k near the k_z-axis is shown in Fig. 4.26 as purple circles. The basis sets for IR T_u are obtained by the action of left coset representatives, i.e. rotations C_3 and C_3^2 about axis (111) on the basis sets (4.14.7), (4.14.8) and (4.14.8). When acting with C_3 on the function $\Psi_3(T_u^{B_{1u}})$ and making use of multiplication table for the elements of O_h group [60] we obtain the functions of the first and second lines:

Table 4.39 Spatial parts of triplet Cooper pairs belonging to IR T_u of T_h group depending on the IR of intermediate group D_{2h} (superscript). Subscripts denote rows of IR T_u

IR	Wavefunction
$\Psi_1\left(T_u^{B_{1u}}\right)$	$\psi_{k_2} + \psi_{C_{2x}k_2} + \psi_{\sigma_y k_2} + \psi_{\sigma_z k_3}$
$\Psi_2\left(T_u^{B_{1u}}\right)$	$\psi_{k_3} + \psi_{C_{2y}k_3} + \psi_{\sigma_z k_3} + \psi_{\sigma_x k_3}$
$\Psi_3\left(T_u^{B_{1u}}\right)$	$\psi_{k_1} + \psi_{C_{2z}k_1} + \psi_{\sigma_x k_1} + \psi_{\sigma_y k_1}$
$\Psi_1\left(T_u^{B_{2u}}\right)$	$\psi_{k_2} - \psi_{C_{2x}k_2} + \psi_{\sigma_y k_2} - \psi_{\sigma_z k_3}$
$\Psi_2\left(T_u^{B_{2u}}\right)$	$\psi_{k_3} - \psi_{C_{2y}k_3} + \psi_{\sigma_z k_3} - \psi_{\sigma_x k_3}$
$\Psi_3\left(T_u^{B_{2u}}\right)$	$\psi_{k} - \psi_{C_{2z}k} + \psi_{\sigma_x k} - \psi_{\sigma_y k}$
$\Psi_1\left(T_u^{B_{3u}}\right)$	$\psi_{k_2} - \psi_{C_{2x}k_2} - \psi_{\sigma_y k_2} + \psi_{\sigma_z k_3}$
$\Psi_2\left(T_u^{B_{3u}}\right)$	$\psi_{k_3} - \psi_{C_{2y}k_3} - \psi_{\sigma_z k_3} + \psi_{\sigma_x k_3}$
$\Psi_3\left(T_u^{B_{3u}}\right)$	$\psi_{k} - \psi_{C_{2z}k} - \psi_{\sigma_x k} + \psi_{\sigma_y k}$

$$\Psi_1(T_u^{B_{1u}}) = C_3\left(\psi_k + \psi_{C_{2z}k} + \psi_{\sigma_x k} + \psi_{\sigma_y k}\right) = \psi_{C_3 k} + \psi_{C_3 C_{2z} k} + \psi_{C_3 \sigma_x k} + \psi_{C_3 \sigma_y k} =$$
$$= \psi_{C_3 k} + \psi_{C_{2x} C_3 k} + \psi_{\sigma_y C_3 k} + \psi_{\sigma_x C_3 k} = \psi_{k2} + \psi_{C_{2x} k2} + \psi_{\sigma_y k2} + \psi_{\sigma_x k2},$$
$$\Psi_2(T_u^{B_{1u}}) = C_3^2\left(\psi_{k_1} + \psi_{C_{2z}k_1} + \psi_{\sigma_x k_1} + \psi_{\sigma_y k_1}\right) = \psi_{C_3^2 k_1} + \psi_{C_3^2 C_{2z} k_1} + \psi_{C_3^2 \sigma_x k_1} + \psi_{C_3^2 \sigma_y k_1} =$$
$$\psi_{C_3^2 k_1} + \psi_{C_{2y} C_3^2 k_1} + \psi_{\sigma_z C_3^2 k_1} + \psi_{\sigma_x C_3^2 k_1} = \psi_3 + \psi_{C_{2y} k_3} + \psi_{\sigma_z k_3} + \psi_{\sigma_x k_3}$$

where $k_2 = C_3 k_1$ and $k_3 = C_3^2 k_1$. Note that the vectors k_1, k_2 and k_3 are directed near the axes k_z, k_x and k_y, respectively (see Fig. 4.26), but in standard form of IR T_u the axes x, y and z correspond to the first, second and third lines, respectively. When acting by the elements of groups C_{2v} with rotational axes C_{2x} and C_{2y} on the functions of the first and of the second lines, respectively, in Table 4.39, it can be obtained that on all coordinate axes function $\Psi\left(T_u^{B_{1u}}\right)$ is nodeless and functions $\Psi\left(T_u^{B_{2u}}\right)$ and $\Psi\left(T_u^{B_{3u}}\right)$ are nodal.

4.15 Some Nontrivial Types of Pairing

Consider a symmorphic space group $I/4mmm$, whose BZ is shown in Fig. 4.5b and its projection on basal plane is shown in Fig. 4.27. The wavefunctions of Cooper pairs in the classical sense of this term are obtained in Sect. 4.5. In this section, we consider some nontrivial pairing states, which are more symmetrical than one-electron states, but these states do not have complete symmetry with respect to the space group.

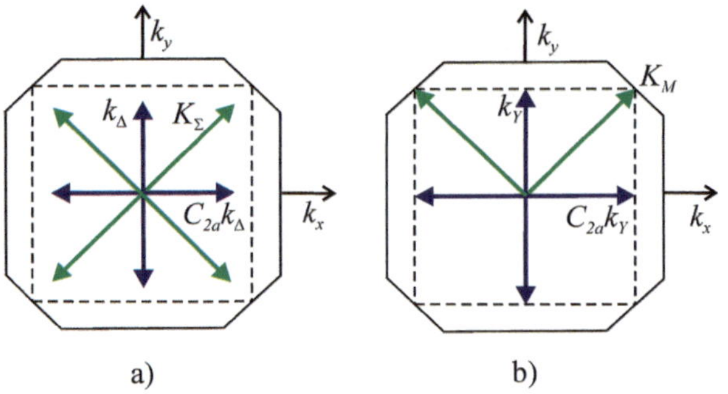

Fig. 4.27 Electron pairs with non-zero total momentum in basal plane of BZ for the group $I4/mmm$. a- electron momentum in direction Δ and pair's momentum in direction Σ, b - electron momentum in point $Y(0, 1/2, 0)$ and pair's momentum in point $M(1/2, 1/2, 0)$.

Begin with the direction k_Δ and consider a Kronecker square $A_{1(\Delta)} \uparrow D_{4h} \times A_{1(\Delta)} \uparrow D_{4h}$, where A_1 is IR of the group of wavevector k_Δ. Figure 4.27a shows the star $\{k_\Delta\}$ of this vector. The group H_Δ of k_Δ is a group C_{2v}, which consists of four elements $\{E, C_{2y}, \sigma_x, \sigma_z\}$. For the double coset, defined by the element $d_\alpha = C_{2a}$, we have

$$k_\Delta + C_{2a}k_\Delta = K_\Sigma(\mu, \mu, 0),$$

where the wavevector of the product is denoted by a capital letter K with subscript.

The star of the wavevector $\{K_\Sigma\}$ consists of four prongs (Fig. 4.27a). The wavevector group H_Σ of K_Σ is also a group C_{2v}, but it consists of different elements $C'_{2v} = \{E, C_{2a}, \sigma_b, \sigma_z\}$. The conjugated group $d_\alpha H_\Delta d_\alpha^{-1}$ is a group C_{2v}, consisting of the elements $\{E, C_{2x}, \sigma_y, \sigma_z\}$. Thus the intersection group in Mackey theorem $M_\alpha = \{E, \sigma_z\}$. When taking the C_{2a} as element a in Mackey theorem we obtain that aM_α=$\{C_{2a}, \sigma_b\}$. Thus the group $\tilde{M}_\alpha$ in Mackey theorem coincides with group C'_{2v} of the wavevector K_Σ. The characters of representations P^+ and P^- appearing in Mackey theorem presented in Table 4.40 correspond to IRs A_1 and B_1 of group C_{2v}, respectively. Thus the symmetrized and antisymmetrized parts of Kronecker square, corresponding to the double coset, defined by C_{2a} are written as $A_{1(\Sigma)} \uparrow D_{4h}$ and $B_{1(\Sigma)} \uparrow D_{4h}$, respectively. It is clear that the result doesn't depend on the label of IR of one-electron state in direction Δ.

Table 4.40 Symmetrized and antisymmetrized squares for $k = Y$, $K = M$

	M		aM		IR of C'_{2v}	IR of D'_{2h}
	E	σ_z	C_{2a}	σ_b		IR
P^+	1	1	1	1	A_1	$A_g + B_{2u}$
P^-	1	1	-1	-1	B_1	$B_{1g} + B_{3u}$

Now consider the point $Y(0, 1/2, 0)$ in $K_\Delta-$direction and double coset, defined by C_{2a}:

$$k_Y + C_{2a}k_Y = K_M(1/2, 1/2, 0).$$

The difference from the previous case is that the group D_{2h} of the wavevector $K_M(1/2, 1/2, 0)$ is larger than the group $\tilde{M}_\alpha$ in Mackey theorem and the induction is performed in two stages. At the first stage, the representations are induced into the point group D_{2h}, and then into the space group D_{4h}. The results of this induction are presented in the last column of Table 4.40. Thus we have for symmetrized square $A_{g(M)} \uparrow D_{4h} + B_{2u(M)} \uparrow D_{4h}$ and for antisymmetrized square $B_{1g(M)} \uparrow D_{4h} + B_{3u} \uparrow D_{4h}$. These states are characterized by parity, but they are not totally symmetric with respect to lattice translations. The star of vector K_Σ consists of two prongs (see Fig. 4.27b) and the elements of the left coset defined by the element C_{2y} interchange the wavefunctions of two prongs. Since the double coset is defined by a rotation, and not by the inversion in the case of classical Cooper pair, there is no direct connection between multiplicity and parity of the pair function.

4.16 Symmetry of Four-Electron States in Solids

It follows clear from the structures of two-electron wavefunctions with zero total momentum that if the interaction exists within the $(k, -k)$ manifold only, the energies of all IRs of the same parity are the same. However, it is usually accepted that the superconductivity in unconventional superconductors is defined by IR of crystal point group. Hence, it follows that we must suppose additional residual interactions between electrons. We suppose that there exists interaction between electrons with almost parallel $k-$vectors. This may take place near planes and lines of symmetry in a BZ. In the former case, two $k-$vectors are connected by mirror reflection operation. In this case, one should consider four-electron states in which $k-$vectors are connected by space inversion I, reflection σ and rotation C_2.

It is convenient to represent symmetry of many-electron states with respect to permutation of electron coordinates by Young tableaux, which represent the IRs of permutation group S_n of n elements. The number of non-equivalent irreducible representations is equal to the number of classes of the group, which is equal to the number of possible partitions of the number n into not-increasing sequence of natural numbers λ_1, $\lambda_2..\lambda_k$ ($\lambda_1 \geqslant \lambda_2 \geqslant \lambda_k > 0$). Therefore, each of the irreducible representations of the symmetric group can be associated with a $[\lambda]$-partition. The partition $[\lambda] = [\lambda_1, \lambda_2...\lambda_k]$ can be visualized as an arrangement of n cells in k rows, with the numbers of cells in successive rows equal to λ_1, $\lambda_2...\lambda_k$. This representation of a partition is called a shape. If the numbers $1...n$ are inserted in these cells we obtain a tableau. The tableau is said to be standard if n natural numbers are placed in such a way that the numbers increase from left to right in all rows and from top to bottom in all columns. The dimension of the IR of group S_n is equal to the number of standard tableaux that can be constructed for a given $[\lambda]-$partition.

Each one of Young tableaux, constructed for a given $[\lambda]$–partition, corresponds to one basis function, having the following property with respect to permutation of two numbers i and j. If i and j are in the same row, the basis function remains unchanged, but if i and j are in different rows, the basis function is multiplied by -1.

For group S_2, there are two partitions [2] and $\left[1^2\right]$ shown in Fig. 4.28a, b, respectively, and for each one there is only one standard placement of natural numbers 1 and 2. Thus both representations are one-dimensional.

The basis sets for Young tableaux of Fig. 4.28a, b are written by two formulas, respectively:

$$\alpha^1\beta^2 + \alpha^2\beta^1 \tag{4.16.1a}$$

$$\alpha^1\beta^2 - \alpha^2\beta^1 \tag{4.16.2a}$$

where α and β denote the wavefunction and superscripts stand for the numbers of electronic coordinates.

The Young tableau $[\bar{\lambda}]$ is called conjugate with respect to the tableau $[\lambda]$, if the lengths of the columns in the table $[\bar{\lambda}]$ equal to the lengths of the rows in the table $[\lambda]$ and vice versa. Two Young tableaux of Fig. 4.28 are mutually conjugate. Total many-electron wavefunctions, represented as direct product of spatial and spin parts, obey the Pauli exclusion principle, i.e. is antisymmetric with respect to the permutation of coordinated of any two electrons. This takes place if spatial and spin parts correspond to conjugate tableaux.

Young tableaux of Fig. 4.28 are mutually conjugate. Thus, in a singlet (triplet) case, permutational symmetry of spatial part corresponds to Young tableau a (b) and spin part corresponds to Young tableau b (a). Thus we can write for singlet and triplet functions in a crystal two following formulas, respectively:

$$\left(k_\alpha^1 k_\beta^2 + k_\alpha^2 k_\beta^1\right) \times \left(s_\zeta^1 s_\eta^2 - s_\zeta^2 s_\eta^1\right) \tag{4.16.3a}$$

$$\left(k_\alpha^1 k_\beta^2 - k_\alpha^2 k_\beta^1\right) \times \left(s_\zeta^1 s_\eta^2 + s_\zeta^2 s_\eta^1\right) \tag{4.16.4}$$

where α and β denote two wavevectors of the same wavevector star and ζ and η correspond to spin projections $s_z = \pm\frac{1}{2}$. The terms $\zeta = 1/2$ and $\eta = -1/2$ appear in singlet and triplet pairs, and there are also terms $\zeta = \eta = 1/2$ and $\zeta = \eta = -1/2$ in a triplet case. For Cooper pairs in spherical symmetry, only space inversion and time reversal are considered and $k_\beta = -k_\alpha$. In a space-group representation theory at a general point of a BZ one electron states are $\left|\bar{G}\right|$–dimensional and Mackey

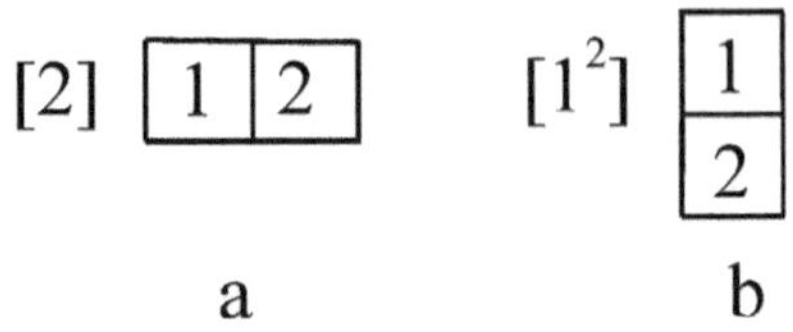

Fig. 4.28 Possible Young tableaux for group S_2

theorem makes it possible to symmetrize the function φ_k with respect to the function φ_{gk} of some other prong. In particular, we can suppose that the wavevectors k_α and k_γ are connected by the reflection σ. If the states with wavevectors k_α and k_γ are close to the symmetry plane (Fig. 4.29) some interaction between them may exist. Also the interaction between the states k and $-k$ should be taken into account. Thus we have four wavefunctions $\{k_\alpha, k_\beta, k_\gamma, k_\delta\}$ connected by I, σ and C_2 (Fig. 4.29a, b).

A complete set of Young tableaux for the group S_4 is shown in Fig. 4.30. Since the group S_4 is isomorphic to the group T_d their IRs are the same. Both groups have two one-dimensional IRs, one two-dimensional IR and two three-dimensional IRs. Since functions of different rows are antisymmetrized and spin space is two-dimensional, the number of rows in spin Young tableau doesn't exceed two and spin tableau symbol is written as $[\lambda_1, \lambda_2]$ in all cases. Since it is possible to antisymmetrize different functions only, the total spin of the state corresponding to a Young table equals to $\frac{1}{2}(\lambda_1 - \lambda_2)$. Thus there are only three possible Young tableaux for spin part of four-electron states, namely, [4] with $S = 2$, [31], with $S = 1$ and [22] with $S = 0$. Corresponding spatial parts belong to conjugate tableaux, i.e. $[1^4]$, [211] and [22], respectively.

Consider as an example the state [22] in a point group symmetry. Let's write the initial function so that the coordinate numbers in superscripts correspond to the numbers of the letters in the subscripts in the Greek alphabet $\{k_\alpha^1, k_\beta^2, k_\gamma^3, k_\delta^4\}$. In these notations, four-electron wavefunctions corresponding to Young tableaux Fig. 4.30c, d may be written as

$$\Psi_c = \left(k_\alpha^1 k_\gamma^3 - k_\alpha^3 k_\gamma^1\right)\left(k_\beta^2 k_\delta^4 - k_\beta^4 k_\delta^2\right) + \left(k_\alpha^2 k_\gamma^4 - k_\alpha^4 k_\gamma^2\right)\left(k_\beta^1 k_\delta^3 - k_\beta^3 k_\delta^1\right) \quad (4.16.5)$$

$$\Psi_d = \left(k_\alpha^1 k_\beta^2 - k_\alpha^2 k_\beta^1\right)\left(k_\gamma^3 k_\delta^4 - k_\gamma^3 k_\delta^4\right) + \left(k_\alpha^3 k_\beta^4 - k_\alpha^4 k_\beta^3\right)\left(k_\gamma^3 k_\delta^4 - k_\gamma^3 k_\delta^4\right) \quad (4.16.6)$$

These two functions form the basis of two-dimensional IR of S_4. Let us suppose that these $k-$vectors correspond to Fig. 4.29. In this case, spatial symmetry of these states consists of inversion I, reflection σ and their product C_2.

Inversion results in simultaneous substitutions $\alpha \to \beta$ and $\gamma \to \delta$. This substitutions results in renumbering of coordinates of Ψ_c without changes of signs. However,

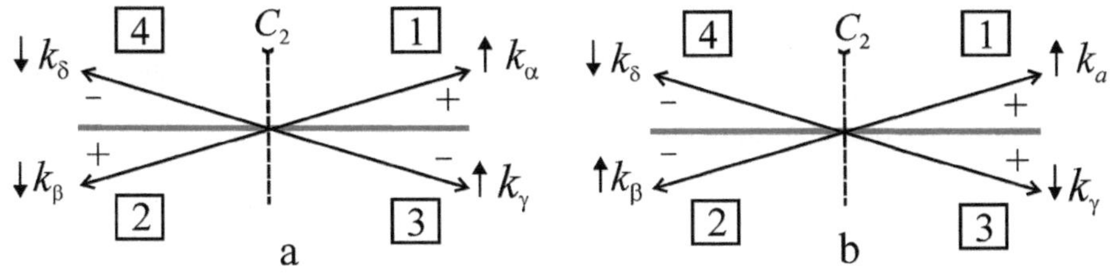

Fig. 4.29 Four-electron states near a symmetry plane. The signs plus and minus correspond to permutational symmetry of spatial parts. For spatial parts, the numbers in the squares in **a** and **b** should be arranged according to Young tableaux presented in Fig. 4.30c, d, respectively

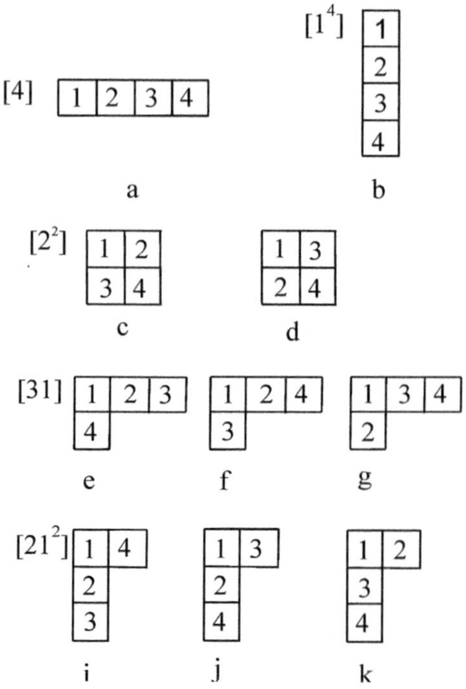

Fig. 4.30 Young tableaux for group S_4

this replacement changes of signs of all multipliers of Ψ_d so the sign of is Ψ_d is also unchanged. The reflection performs substitutions $\alpha \to \gamma$ and $\beta \longrightarrow \delta$, resulting in interchanging the signs of all multipliers of Ψ_c and renumbering the coordinates in Ψ_d. Thus both functions are invariant under the elements of group C_{2h}. However, the functions Ψ_c and Ψ_d have nontrivial symmetry with respect to permutations of coordinates of two electrons. Function Ψ_c may be considered as constructed from two even (singlet) pairs and the signs of one-electron functions change under the action of reflection. Similarly the function Ψ_d is constructed from two odd (triplet) pairs and the signs of one-electron wavefunctions are unchanged under the action of reflection. These signs are shown explicitly in Fig. 4.29.

Thus we obtained that four-electron states corresponding to Young tableau [2, 2] constructed from singlet or triplet pairs have total spin $S = 0$. Also on the basis of triplet two-electron functions one can construct by the similar way the states with $S = 2$. In this case, spin part corresponds to Young tableau [4] and spatial part corresponds to Young tableau $[1^4]$.

Bibliography

1. M. Dresselhaus, G. Dresselhaus, A. Jorio, *Group Theory Application to the Physics of Condensed Matter* (Springer, 2008)
2. M. Hamermesh, *Group Theory and Its Application to Physical Problems* (Adison-Wesley, New York, 1964)
3. M.I. Petrashen, E.D. Trifonov, *Applications of Group Theory in Quantum Mechanics* (MIT Press, 1969)
4. I.G. Kaplan, *Symmetry of Many-Electron Systems* (Academic Press, New York, 1975)
5. E.P. Wigner, *Group Theory and its Application to the Quantum Mechanics of Atomic Spectra* (N Y Academic Press, 1959)
6. J.L. Birman, *Theory of Crystal Space Groups and Lattice Dynamics* (Springer, 1984)
7. R.S. Knox, A. Gold, *Symmetry in the Solid State* (W. A. Benjamin, N Y, 1964)
8. H.W.Streitwolf, *Gruppentheorie in der Festkorperphysik* (Leipzig, 1967)
9. C.J. Ballhausen, *Introduction to Ligand Field Theory* (McGraw-Hill, N.Y., 1962)
10. I.B. Bersuker, *Electronic Structure and Properties of Transition Metal Compounds: Introduction to the Theory*, 2nd edn. (Wiley, 2010)
11. F.A. Cotton, *Chemical Applications of Group Theory*, 3edn. (Wiley, 1990)
12. A. Streitwieser, *Molecular Orbital Theory for Organic Chemists* (Wiley, New York, 1961)
13. C. Chisholm, *Group Theoretical Techniques in Quantum Chemistry* (Academic Press, London; N. Y., 1976)
14. A.P. Cracknell, *Applied Group Theory* (Pergamon Press, 1968)
15. L.D. Landau, E.M. Lifshitz, *Quantum Mechanics: Non-relativistic Theory* (Pergamon, 1965)
16. V.P. Smirnov, R.A. Evarestov, Application of the site symmetry for the simplification of the group theoretical analysis of electronic and vibrational states of crystals. Bull. Leningrad State Univ. **4**, 107 (1981)
17. R.A. Evarestov, V.P. Smirnov, *Site Symmetry in Crystals*. Springer Series in Solid-State Sciences, vol. 108 (1993)
18. V.G. Iarzhemskii, E.N. Muraviev, Method of induced representation in the theory of molecular orbitals. Doklady Akademii Nauk SSSR **278**, 945 (1984)
19. C.J. Bradley, A.P. Cracknell, *The Mathematical Theory of Symmetry in Solids* (Representation Theory for Point Groups and Space Groups (Clarendon, Oxford, 1972)
20. S.L. Altmann, *Induced Representations in Crystals and Molecules* (Academic Press, London, 1977)

V. G. Yarzhemsky, *Novel Group Theoretical Methods for Electron Structure Theory*,
https://doi.org/10.1007/978-3-031-72802-0

21. S.L. Altmann, P. Herzig, *Point-Group Theory Tables* (Clarendon Press, 1994)
22. A.J. Coleman, *Induced and Subduced Representations*, in book *Group Theory and its Applications*, ed. by E.M. Loebl (Academic Press, 1968), pp. 57–118
23. W. Lederman, *Introduction to Group Characters* (Cambridge University Press, 1987)
24. R.W. Haase, P.H. Buttler, Coupling, subduction and induction transformations for group representations. J. Phys. A: Math. Gen. **17**, 47 (1984)
25. R.W. Haase, R. Dirl, The induction coefficient of induced representation theory: the algebra and analytical expression. J. Phys. A: Math. Gen. **20**, 1065 (1987)
26. V.G. Yarzhemsky V.I. Nefedov, Group theoretical description of two-electron wave functions in systems with subgroups of symmetry. Int. J. Quant. Chem. **100**, 519 (2004)
27. V.G. Yarzhemsky, Internal symmetries and additional quantum numbers for nanoparticles. Nanoscale Syst. Math. Model. Theory Appl. **2**, 96 (2013)
28. V.G. Yarzhemsky, A.D. Izotov, V.O. Izotova, Additional quantum numbers for vibration states of symmetric nanoparticles. Doklady Phys. **58**(12), 524 (2013)
29. V.G. Yarzhemsky, E.N. Murav'ev, Orbits and induced representations in the quantum chemistry of nanostructures Russ. J. Inorg. Chem. **54**, 1273 (2008)
30. V.G. Yarzhemsky, Induced representation method in the theory of electron structure and superconductivity. Adv. Math. Phys. 4873914 (2019)
31. G.F. Koster, Localized functions in molecules and crystals. Phys. Rev. **89**, 67 (1953)
32. J. des Cloizeaux, Orthogonal orbitals and generalized Wannier functions. Phys. Rev. **129**, 554 (1963)
33. J. Zak, Band representations of space groups. Phys. Rev. B. **26**, 3011 (1982)
34. R.A. Evarestov, V.P. Smirnov, *Site Symmetry in Crystals* (Springer, 1997)
35. G.W. Mackey, On induced representations of groups. Am. J. Math. **73**, 576 (1951)
36. G.W. Mackey, Symmetric and anti symmetric Kronecker squares and intertwining numbers of induced representations of finite groups. Am. J. Math. **75**, 387 (1953)
37. C.J. Bradley, Space groups and selection rules. J. Math. Phys. **7**, 1145 (1966)
38. C.J. Bradley, B.L. Davies, Kronecker products and symmetrized squares of irreducible representations of space groups. J. Math. Phys. **11**, 1536 (1970)
39. V.G. Yarzhemsky, Mackey theorem and two-electron wave function of a multi-centre system. Few-Body Syst. **22**, 27 (1997)
40. V.G. Yarzhemsky, E.N. Murav'ev, Space group approach to the wavefunction of a Cooper pair. J. Phys. Cond. Matter **4**, 3525 (1992)
41. V.G. Yarzhemsky, Space-group approach to the nodal structure of the superconducting order parameter in UPt_3. Physica Status Solidi (b) **209**, 101 (1998)
42. T. Micklitz, M.R. Norman, Odd parity and line nodes in nonsymmorphic superconductors. Phys. Rev. B **80**, 100506(R) (2009)
43. T. Micklitz, M.R. Norman, Nodal lines and nodal loops in nonsymmorphic odd-parity superconductors. Phys. Rev. B **95**, 024508 (2017)
44. T. Nomoto, H. Ikeda, Symmetry-protected line nodes in non-symmorphic magnetic space groups: applications to UCoGe and UPd_2Al_3. J. Phys. Soc. Jap. **86**, 023703 (2017)
45. S. Sumita, Y. Yanase, Unconventional superconducting gap structure protected by space group symmetry. Phys. Rev. B **97**, 134512 (2018)
46. S. Sumita, T. Nomoto, K. Shiozaki, Y. Yanase, Classification of topological crystalline superconducting nodes on high-symmetry lines: point nodes, line nodes, and Bogoliubov Fermi surfaces. Phys. Rev. B **99**, 134513 (2019)
47. S. Sumita, *Modern Classification Theory of Superconducting Gap*, Springer Theses Book Series (2021)
48. K.S. Chan, D.J. Newman, The unique labelling of states via Mackey's theorem. J. Phys. A: Math. Gen. **17**, 253 (1984)
49. V.G. Yarzhemsky, E.A. Teplyakov, Time reversal symmetry and the structure of Cooper pair wavefunction in topological superconductor UTe_2. Phys. Lett A. **384**, 126724 (2020)
50. V.G. Yarzhemsky, Multiplicity, parity and angular momentum of a Cooper pair in unconventional superconductors of D_{4h} symmetry: Sr_2RuO_4 and Fe-pnictide materials. Symmetry **13**, 1435 (2021)

51. V.G. Yarzhemsky, E.A. Teplyakov, Additional quantum numbers for two-electron states in solids. Application to topological superconductor UPt_3. J. Phys. A Math. Theor. **54**, 455304 (2021)
52. V.G. Yarzhemsky, *Symmetry of Two and Four-Electron States in Solids*, in *book Symmetry, Spectroscopy and SCHUR*, ed. by C. King, M. Bylicki, J. Karwowski, Torun (Poland), pp. 343–353 (2006). ISBN 83-231-1901-5
53. V.G. Yarzhemsky, Symmetry of Two- and Four- Electron States in Solids. Application to Unconventional Superconductors, arXiv:cond-mat/0510233
54. F.J. Dyson, Application of group theory in particle physics. SIAM Rev. **8**, 1 (1966)
55. E.P. Wigner, The unreasonable effectiveness of mathematics in the natural sciences. Commun. Pure Appl. Math. **13**, 001–014 (1960)
56. S.L. Altman, Double groups and projective representations. Mol. Phys. **38**, 489 (1979)
57. W.G. Harter, Algebraic theory of ray representations of finite groups. J. Math. Phys. **10**, 739 (1969)
58. A.P. Cracknell, Group theory in solid-state physics is not dead yet alias some recent developments in the use of group theory in solid-state physics. Adv. Phys. **23**, 673 (1974)
59. L. Pau1ing, *The Nature of the Chemical Bond*, 3rd edn. (Cornell University Press, Ithaca, NY, 1960), pp. 108–182
60. O.V. Kovalev, *Irreducible Representations of the Crystallographic Space Groups: Irreducible Representations, Induced Representations and Corepresentations* (Gordon & Breach, N.Y., 1993)
61. N.B. Melnikov, B.I. Reser, *Space Group Representations* (Theory, Tables and Applications (Springer, 2022)
62. L.P. Bouckaert, R. Smoluchowski, E. Wigner, Theory of Brillouin zones and symmetry properties of wave functions in crystals. Phys. Rev. **50**, 58 (1936)
63. G. Calzaferri, R. Rytz, The band structure of diamond. J. Phys. Chem. **100**, 11122 (1996)
64. E.P. Wigner, Symmetry principles in old and new physics. Bull. Am. Math. Soc. **74**, 793 (1968)
65. J. Bardeen, L.N. Cooper, I.R. Schriffer, Theory of superconductivity. Phys. Rev. **108**, 3259 (1957)
66. M.R. Norman, The challenge of unconventional superconductivity. Science **332**, 196 (2011)
67. D.J. Bishop, C.M. Varma, B. Batlogg, E. Bucher, Z. Fisk, J.L. Smith, Ultrasonic attenuation in UPt_3. Phys. Rev. Lett. **53**, 1009 (1984)
68. J.D. Strand, D.J. Bahr, D.J. Van Harlingen, J.P. Davis, W.J. Gannon, W.P. Halperin, The transition between real and complex superconducting order parameter phases in UPt_3. Science **328**, 1368 (2010)
69. E.R. Schemm, W.J. Gannon, C.M. Wishne, W.P. Halperin, A. Kapitulnik, Observation of broken time-reversal symmetry in the heavy fermion superconductor UPt_3. Science **345**, 190 (2014)
70. Y. Maeno, H. Hashimoto, K. Yoshida, S. Nishizaki, T. Fujita, J.D. Bednorz, F. Lichtenberg, Superconductivity in a layered perovskite without copper. Nature **372**, 532 (1994)
71. Y. Maeno, S. Kittaka, T. Nomura, S. Yonezawa, K. Ishida, Evaluation of spin-triplet superconductivity in Sr_2RuO_4. J. Phys. Soc. Jpn. **81**, 011009 (2012)
72. A. Pustogow, Y. Luo, A. Chronister, Y.S. Su, D.A. Sokolov, F. Jerzembeck, A.P. Mackenzie, C.W. Hicks, N. Kikugawa, S. Raghu, E.D. Bauer, S.E. Brown, Constraints on the superconducting order parameter in Sr_2RuO_4 from oxygen-17 nuclear magnetic resonance. Nature **574**, 72 (2019)
73. V. Grinenko, S. Ghosh, R. Sarkar, J.-C. Orain, A. Nikitin, M. Elender, D. Das, Z. Guguchia, F. Brückner, M.E. Barber, J. Park, N. Kikugawa, D.A. Sokolov, J.S. Bobowski, T. Miyoshi, Y. Maeno, A.P. Mackenzie, H. Luetkens, C.W. Hicks, H.-H. Klauss, Split superconducting and time-reversal symmetry-breaking transitions in Sr_2RuO_4 under stress. Nat. Phys. **17**, 748 (2021)
74. S. Ran, C. Eckberg, Q.-P. Ding, Y. Furukawa, T. Metz, S.R. Saha, I.-L. Liu, M. Zic, H. Kim, J. Paglione, N.P. Butch, Nearly ferromagnetic spin-triplet superconductivity. Science **365**, 684 (2019)

75. S. Ran, I.-L. Liu, Y.S. Eo, D.J. Campbell, P.M. Neves, W.T. Fuhrman, S.R. Saha, C. Eckberg, H. Kim, D. Graf, F. Balakirev, J. Singleton, J. Paglione, N.P. Butch, Extreme magnetic field-boosted superconductivity. Nat. Phys. **15**, 1250 (2019)
76. M. Sato, Y. Ando, Topological superconductors: a review. Rep. Prog. Phys. **80**, 076501 (2017)
77. Y. Tanaka, M. Sato, N. Nagaosa, Symmetry and topology in superconductors -odd-frequency pairing and edge states. J. Phys. Soc. Jpn. **81**, 011013 (2012)
78. P.W. Anderson, Structure of "triplet" superconducting energy gap. Phys. Rev. B **30**, 4000 (1984)
79. G.E. Volovik, L.P. Gorkov, Superconducting classes in heavy-fermion systems. Sov. Phys. JETP **61**, 843 (1985)
80. T.M. Rice, M. Sigrist, Sr_2RuO_4: an electronic analogue of ^3He? J. Phys. Condens. Matter. **7**, L643 (1995)
81. K. Ueda, T.M. Rice, p-wave superconductivity in cubic metals. Phys. Rev. B **51**, 7119 (1985)
82. V.P. Mineev, K.V. Samokhin, *Introduction to Unconventional Superconductivity* (N.Y Gordon and Breach, 1999)
83. J.A. Sauls, The order parameter for the suprconduction phases of UPt_3. Adv. Phys. **43**, 113 (1994)
84. K. Deguchi, Z.Q. Mao, Y.J. Maeno, Determination of the superconducting gap structure in all bands of the spin-triplet superconductor Sr_2RuO_4. J. Phys. Soc. Jpn. **73**, 1313 (2004)
85. D. Aoki, J.-P. Brison, J. Flouquet, K. Ishida, G. Knebel, Y. Tokunaga, Y. Yanase, Unconventional superconductivity in UTe_2. J. Phys.: Condens. Matter **34**, 243002 (2022)
86. V.G. Yarzhemsky, Wavefunction of a CooperPairinCrystals of D^{1}_{2h} and D^{17}_{4h} symmetry Zeitsch. Phys. B: Cond. Matter **99**, 19 (1995)
87. V.G. Yarzhemsky, Space-group approach to the nodal structure of superconducting order parameter in ferromagnetic and antiferromagnetic materials. Int. J. Quant. Chem. **80**, 133 (2000)
88. V.G. Yarzhemsky, *Space-Group Approach to the Wavefunction of a Cooper Pair and its Application to Hight-Tc Superconductors*, in Lectures on the Physics of Highly Correlated Electron Systems, AIP Conf. Proc. **678**, 343 (Melville, New York, 2003)
89. V.G. Yarzhemsky, V.I. Nefedov, Group theoretical treatment of photoelectron spectra of high-T_c superconductors. Hidden symmetry and colour pairs. Phil. Mag. Lett. **86**(11), 733-7 (2006)
90. V.G. Yarzhemsky, Space-group approach to two-electron states in unconventional superconductors. J. Opt. Adv. Mater. **10**, 1759 (2008)
91. V.G. Yarzhemsky, V.I. Nefedov, Time-reversal symmetry violation and the structure of the superconducting order parameter of $PrOs_4Sb_{12}$. Phys. Solid State **51**, 448 (2009)
92. V.G. Yarzhemsky, Nodal quantum numbers for two-electron states in solids. Few-Body Syst. **53**, 499 (2012)
93. V.G. Yarzhemsky, Group theoretical lines of nodes in triplet chiral superconductor Sr_2RuO_4. J. Phys. Soc. Jpn. **87**, 114711 (2018)
94. V.G. Yarzhemsky, E.A. Teplyakov, Topological structure of the order parameter of unconventional superconductors based on d- and f-elements. Symmetry **15**, 376 (2023)
95. E.I. Blount, Symmetry properties of triplet superconductors. Phys. Rev. B **32**, 2935 (1985)
96. S. Kobayashi, K. Shiozaki, Y. Tanaka, M. Sato, Topological Blount's theorem of odd-parity superconductors. Phys. Rev. B **90**, 024516 (2014)
97. K. Izawa, Y. Nakajima, J. Goryo, Y. Matsuda, S. Osaki, H. Sugawara, H. Sato, P. Thalmeier, K. Maki, Multiple superconducting phases in new heavy fermion superconductor $PrOs_4Sb_{12}$. Phys. Rev. Lett. **90**, 117001 (2003)

GPSR Compliance

The European Union's (EU) General Product Safety Regulation (GPSR) is a set of rules that requires consumer products to be safe and our obligations to ensure this.

If you have any concerns about our products, you can contact us on ProductSafety@springernature.com

In case Publisher is established outside the EU, the EU authorized representative is:

Springer Nature Customer Service Center GmbH
Europaplatz 3
69115 Heidelberg, Germany

Batch number: 10732044

Printed by Printforce, the Netherlands